Crucial Experiments in Modern Physics

George L. Trigg

Crane, Russak & Company, Inc.
NEW YORK

Crucial Experiments
in Modern Physics

Published in the United States by
Crane, Russak & Company, Inc.
347 Madison Avenue
New York, N. Y. 10017

Previously Published as
Van Nostrand Reinhold Momentum Book No. 23

LC: 75–21567

ISBN: 0–8448–0765–6

Printed in the United States of America

To the memory of
George Eric Macdonnell Jauncey

Preface

In the summer of 1965, a conference on New Instructional Methods in Physics was held at the University of Washington, Seattle. At a preliminary planning discussion, the point was raised that many students have difficulty grasping some of the concepts of modern physics, especially those of quantum theory, and that a recounting of the experimental bases for these concepts would be useful. This work is an attempt to meet that need.

The choice of what experiments to include was necessarily somewhat arbitrary. It was clear that heavy emphasis should be placed on the ideas of quantum theory, as they are so often in conflict with "common sense." On the other hand, the special theory of relativity was not included despite its similar conflict with "common sense," first because it stems almost completely from a single experiment that is too well known to need repeating in this context, and second because it is adequately dealt with by Robert Katz in *An Introduction to the Special Theory of Relativity*, MOMENTUM Book No. 9. Similarly, studies of the properties of the electron are of unquestioned importance, but are already included in David L. Anderson's *The Discovery of the Electron*, MOMENTUM Book No. 3.

This book can be used at several levels. It is designed to be intelligible to a college sophomore or advanced freshman who has at least a reasonable qualitative grasp of classical physics; for him it should facilitate acceptance of the sometimes paradoxical ideas at the base of modern physics. It should also be useful to the advanced student who wishes to place his knowledge in a better historical perspective. To this end, I have attempted to present each experiment—with one exception, which is explicitly mentioned—from the point of view of the average physicist of the time. I have tried to make the discussion of the experiments reasonably complete in that the essentials of the method and the difficulties both of execution and of interpretation are made clear.

Most of the references to the original works were obtained from two sources. One is J. D. Stranathan, *The "Particles" of Modern Physics* (The Blakiston Company, Philadelphia, 1942), which has served as a reference source to a generation of physicists. The other is *The World of the Atom,* edited by Henry A. Boorse and Lloyd Motz (Basic Books, Inc., New York, 1966). Many of the original papers are republished or translated there, in whole or in part, and are cited in the chapter bibliographies of this book merely by the title of the collection and a page. The translations of quotations used here, however, are my own.

Many people have contributed to the making of this book. I am grateful to the Commission on College Physics, the University of Washington, and the National Science Foundation for support during the Seattle Conference, and to the staff of the Conference for efficient and cheerful service. I thank Brookhaven National Laboratory for its hospitality; particular mention must be made of the Illustration Division, who provided several of the original drawings and assisted with most of the reproduced figures, and of the Research Library, especially Mrs. Marilyn Galli, who helped locate and provide much source material. I have benefited greatly from critical readings of a preliminary version by Dr. Larry Wilets, Dr. Robert Weinstock, and Mr. Roger Atlas, and of the final version by Miss Anne Young. An admirable job of typing was done by Mrs. Jean Walker, Mrs. Regina Scalzo, and especially Miss Veronica Oertel. Finally, I acknowledge my debt to Dr. Walter Michels, who, first as a member of the Commission on College Physics, later as General Editor of the MOMENTUM series, and always as a friend, has helped at every stage of the work and has provided continual advice and encouragement.

George L. Trigg

Upton, New York

Acknowledgments

I am indebted to the following sources for permission:

To the Deutsche Physikalische Gesellschaft, for translated quotations and figures from papers by Lummer and Pringsheim, by Planck, and by Franck and Hertz, from *Verhandlungen der Deutschen Physikalischen Gesellschaft.*

To George Allen & Unwin, Ltd., and Wiley Interscience, Inc., for quotations and figures from *The Collected Papers of Lord Rutherford of Nelson,* edited by Sir James Chadwick (George Allen & Unwin, Ltd., London).

To Mrs. Muriel Howorth, for quotations from *Pioneer Research on the Atom: The Life Story of Frederick Soddy,* by Muriel Howorth (New World Publications, Eastbourne, Sussex, England).

To the Centre National de la Recherche Scientifique, Republique Française, for translated quotations and figures from *Oeuvres Scientifiques,* by Jean Perrin (Service des Publications de la C.N.R.S., Paris).

To the Royal Society, for quotations and figures from papers by Geiger and by Geiger and Marsden, from *Proceedings of the Royal Society.*

To Taylor and Francis, Ltd., for quotations and figures from a paper by Geiger and Marsden, from *The Philosophical Magazine.*

To Cambridge University Press, for quotations from *Backgrounds to Modern Science,* edited by Joseph Needham and Walter Pagel (Cambridge University Press, London).

To The American Physical Society, for quotations and figures from papers by Millikan, by Compton, and by Davisson and Germer, from *The Physical Review.*

To Springer-Verlag, for translated quotations and figures from papers by Stern, by Stern and Gerlach, by Knauer and Stern, by Estermann and Stern, and by Estermann, Frisch, and Stern, from *Zeitschrift für Physik.*

To Dr. L. H. Germer, for quotations and figures from his paper with Davisson, from *The Physical Review.*

Contents

	Preface	v
1	Introduction	1
2	The Origin of the Quantum Concept	4
3	Transmutation of Elements	23
4	The Existence of Atoms	37
5	The Atomic Nucleus	55
6	Collisions of Electrons with Atoms	69
7	The Photoelectric Effect	76
8	Spatial Orientation of Atomic Magnets	88
9	Particle Properties of Light	97
10	Wave Properties of Matter	105
	Appendix A	135
	Appendix B	137
	Index	139

1 *Introduction*

To the student of today, the term "modern" may seem a strange one to apply to a group of experiments most of which were performed before 1930. Nevertheless, this term, inadequate as it is, has become the one conventionally used to designate one of the two major historical divisions of physics. The "classical" period, which began with Galileo and Newton and ended about 1900, had seen the development of dynamics into an almost closed branch of physics that served as an ideal for the other branches; had seen, in little more than a single century's time, electromagnetism grow from a set of parlor tricks to an elegant theoretical structure that encompassed most of optics as well; and had seen a similar transformation of the empirical arts of working with heat into the science of thermodynamics. In sum, by the end of the nineteenth century the physicist had a reasonably consistent and coherent, and fairly complete, understanding of the facts of the physical universe as he knew them. To be sure, there were some flaws and gaps; but there seemed to be little reason to doubt that these would be remedied by measuring "the next place of decimals." [1]

This confidence proved to be misplaced. Beginning about 1895, experimental research more and more produced findings which were completely out of accord with the classical ideas. In some cases only a fairly easy readjustment was enough. In others, a radical alteration of ideas was necessary, and the situation cannot be regarded as completely settled even today. The experimental evidence actually, of course, consists of many individual pieces, forming a structure much of whose strength arises from the interconnecting relations among the pieces. But it is possible, for many of the most basic points, to single out individual works of which it can be said, "This was the decisive result." The experiments to be treated in the following chapters are of this sort. In order to appreciate fully their significance and impact, we must first note some characteristic features of classical physics.

In the first place, its concepts had been, without exception, developed in the range of ordinary experience and extended as needed. This is not surprising; it is natural human behavior to expect that regularities which hold under a fairly wide range of conditions will continue to hold under all conditions. Thus it would be futile to send a rocket to Mars carrying equipment with which to photograph the surface of that planet unless we were reasonably confident that the equipment, and the rocket itself, would behave near Mars in accordance with the same laws that we have discovered on the surface of the earth. Similarly, we have faith that an automobile designed five years ago in the light of laws as they were known then will still operate properly tomorrow.

But such extensions, or *extrapolations,* can be misleading. For example, a famous law of physics, discovered in 1662 by Robert Boyle, states that the product of the volume of a particular quantity of gas and the pressure it exerts on the walls of its container is constant as long as the temperature remains unchanged, any increase in one of the two quantities being exactly compensated by a decrease in the other and vice versa. If the temperature is too low, however, or the pressure too high, the law fails; the gas starts to condense, and there may be a substantial change of volume with no compensating change of pressure. Such a failure of an extrapolation is commonly linked with the existence of laws or phenomena not envisioned in the original relationship. In our example, the failure of Boyle's law is related to the fact that gases can condense to form liquids.

The second point of concern to our discussion has to do with the classical picture of the constitution of matter. An attitude that harked back to Aristotle admitted only two possibilities: Matter could be continuous, in analogy to gelatin, or it could have a discrete structure like a pile of rocks. Classical theory had developed according to the first alternative. To be sure, the atomic hypothesis was well known; but even among chemists, where its staunchest supporters might be expected to have been found, there was strong and outspoken opposition. Apart from the beginnings of what is now known as kinetic theory, the atomic concept had little role in classical physics. Where it did enter, it was a rudimentary version, with little or no concern over possible structure to the atom.[2]

This, then, is the background against which the work presented here is to be viewed. Some parts of the work will stand clearly distinct from the background; others will appear to be only parts of the background viewed in a new light. The unifying feature of the work is that all the experiments to be described deal with phenomena which, either in scale or in subtlety, are beyond the range of our senses and therefore outside our ordinary experience. The key to an appreciation of them may be described, therefore, as a refusal to be restrained by "common sense" outside of the realm where common sense has been developed.

FOOTNOTES

[1] It is ironic that this expression, apparently originally due to Lord Kelvin, was given widespread circulation through its use in the catalogue of the University of Chicago by A. A. Michelson, whose own experimental work was to provide the basis for one of the two great revolutions of thought that marked the first half of the twentieth century in physics.

[2] There was little cause for such concern. After all, it was not until 1897 that the electron was discovered introducing the possibility of an internal structure and subdivision of the *atomos*—the indivisible.

2 *The Origin of the Quantum Concept*

Two great conceptual structures have characterized the development of physics during the first half of the twentieth century: relativity and quantum theory. Of the two, the former is generally regarded as being founded on a nineteenth century experiment, the Michelson-Morley experiment; the theory sprang, so to speak, essentially complete from the mind of Albert Einstein,[1] and it will not be further treated in this work.

Quantum theory has a much more involved history, and in fact most of the experiments treated in this book pertain to it. It has come to be thought of primarily as a theory of the structure of atoms and aggregates of atoms. It is of considerably broader significance, however; and it had its origin in an attempt to give a complete description of the radiation from a small hole in the wall of a furnace. For reasons that will be explained, such radiation had been the subject of extensive theoretical studies. Some general properties had been deduced, but the only explicit formula rested on questionable assumptions. This chapter will describe the first experimental work that was both accurate enough and extensive enough to test the formula. The formula proved inadequate; the necessary modification of it contained a new constant of nature, which became known as the "quantum of action."

In order to discuss these developments, it is necessary to introduce an idealized object called a black body. It is to be recalled that when light strikes the surface of an object, two effects occur:

Some of the light is *reflected* from the surface, and some passes through the surface into the body of the object. The latter portion, in turn, undergoes at least one and possibly two further processes: Some (or all) of it is *absorbed;* some may reach another surface and pass out, or be *transmitted.* We see an object, unless it is intrinsically luminous, only because it reflects some light into our eyes. The less light an object reflects, the darker it appears to be. If an object existed that absorbed all the light that fell on it, it would reflect none, and would appear perfectly black.[2] An object of this sort is called an ideal black body. No such body exists. Nevertheless, it is quite possible—and useful—to act as though it did, and to determine many of the properties it would have; this sort of idealization is quite common in science, and is especially valuable when, as will be seen to be true in the present case, the ideal situation can be closely approximated in reality.

Of course, the radiation that is absorbed by a black body carries energy. The internal energy of the body would be increased and the temperature of the body would rise indefinitely if there were no mechanism by which the body disposed of some energy. There must be, in fact, a mechanism that is available even to a body isolated in a vacuum, where conduction and convection are impossible. It is, simply, that the body radiates; in fact, not only a black body but any object left to itself in unchanging surroundings tends toward a state in which it radiates as much energy per unit time—as much power—as it absorbs.[3]

The wavelengths at which energy is radiated range continuously over the spectrum, including not only the visible but also the ultraviolet, the infrared, and all other portions. The way in which the energy is distributed among the various wavelengths changes with the temperature of the body,[4] as well as being dependent on the material of the surface of the body. The measure of this distribution in wavelength is called the *spectral emittance,* designated by E_λ and defined as follows: The spectral emittance at wavelength λ is the energy associated with wavelengths around λ radiated per unit time, per unit surface area, per unit wavelength. Thus, $E_\lambda \, d\lambda \, dS \, dt$ is the energy in the wavelength range between λ and $\lambda + d\lambda$ radiated in time dt by an element of surface of area dS. The spectral emittance of a *black* body, however, does *not* depend on the properties of any particular substance, and so has a sort of universal or absolute character. This is one reason for interest in a black body. Because of its special importance, we will denote the spectral emittance of a black body by a special symbol, e_λ.

Let us consider the steps that made blackbody radiation amenable to study. We will consider a cavity—a region of space bounded by material of any kind. Suppose the walls of the cavity are maintained at a uniform temperature T. The cavity will certainly contain some radiation, because the walls will radiate. G. R. Kirchhoff, in introducing the concept of a black body in 1860, proved that the radiation has the following significant properties: First, it is isotropic (the same in any direction) and homogeneous (the same at every point in the cavity), not only over all but at any specified wavelength. Second, it is identical with the radiation emitted by a black body, in the sense that the *irradiancy,* the energy striking unit area per unit time, is equal to the energy emitted by unit area of a black body per unit time; and this also is true at any wavelength as well as over all. The isotropy and homogeneity are proved by establishing that otherwise it would be possible to produce perpetual motion; thus, for example, if the radiation were more intense at one point than another, identical absorbers placed at the two points would come to different temperatures, and could be used one as source and the other as sink for a heat engine that would extract work from the system without producing any other change.

To establish the equivalence between cavity radiation and blackbody radiation, suppose that a black body at temperature T is placed in a cavity whose walls are also at temperature T. The energy falling on unit area of the black body per unit wavelength around λ per unit time we will denote by R_λ; by definition of a black body, it is all absorbed. The energy emitted per unit area per unit wavelength per unit time is just e_λ. These energies must balance, or else we would have spontaneous transfer of energy between two objects at the same temperature (the possibility of an imbalance at one wavelength being canceled by an opposite imbalance at another can be ruled out by arguments involving auxiliary apparatus such as selective reflectors); in symbols,

$$R_\lambda = e_\lambda.$$

As we shall see, the possibility of experimental measurement of e_λ rests essentially on this identity.

Next, suppose that a nonblack body at temperature T is placed in the cavity. The incident power per unit area per unit wavelength is still R_λ; but now not all of it is absorbed. The fraction absorbed at wavelength λ is called the *absorptance* and denoted by a_λ; it may depend on the temperature as well as the material of the surface. The power absorbed per unit area per unit wavelength around λ, then, is $a_\lambda R_\lambda$. On the other hand, the power emitted per unit area per unit wavelength around λ is E_λ. Again the two must balance, so that

$$a_\lambda R_\lambda = E_\lambda,$$

or

$$E_\lambda / a_\lambda = R_\lambda.$$

By virtue of the earlier result, we have

$$E_\lambda / a_\lambda = e_\lambda:$$

The ratio between spectral emittance and absorptance at a given wavelength and temperature is *independent* of the material involved.

This relationship gives blackbody radiation an added significance. O. Lummer and E. Pringsheim, whose experimental studies of blackbody radiation are to be described in this chapter, put it as follows:

> If one knows, therefore, the radiation of a black body as a function of temperature, then one thereby knows the laws of radiation for all bodies whose absorptive power is also given as a function of wavelength and temperature. Experimentally, the reverse process is likely to be simpler, to explain the absorption A with the help of the knowledge of e by the study of the radiation of the body.

These considerations had already caused a fair amount of attention to be devoted to the problem of blackbody radiation by the end of the nineteeth century. Most of the efforts were empirical deductions from observations on the spectral distributions of radiation from real bodies. In 1896, F. Paschen, reporting on his own results of this type, cited some half-dozen earlier works; Paschen's expression, $E_\lambda = C\lambda^{-a}e^{-c/\lambda T}$, where T is the absolute temperature and C, a, and c are constants determined by the material, was one of the simplest.[5] The one empirical result that has retained its validity and hence its usefulness is one that was given in 1879 by J. Stefan, stating that the *total* emittance, the total power (including all wavelengths) radiated per unit area, is proportional to the fourth power of the absolute temperature of the radiating body.

Theoretical studies were not lacking. In 1884, Ludwig Boltzmann published two papers which together provided a proof that Stefan's empirical relationship must hold for a black body. (As a consequence, the relationship has become known as the Stefan-Boltzmann law.) Nine years later, Willy Wien obtained two equally important results. He noted that if a cavity were reduced in volume, say by moving a piston, the energy per unit volume of the radiation in the cavity would be increased—not only because the energy already present was confined to a smaller volume, but also because the radiation exerts pressure on the piston, and work is done against that pressure in the act of compression. The energy density can also be increased by an increase in the temperature of the cavity, and the second law of thermodynamics relates the increases produced in the two ways. The relationship must hold not only over all, but also for the energy density associated with every infinitesimal range of wavelengths. But the moving walls produce a change in wavelength of the reflected radiation, by the Doppler effect; and so the temperature change must also alter the wavelength distribution. The quantitative consequences of this were twofold. First, if one is to make comparisons of things dependent on wavelength at two different temperatures, it is not proper to compare values at the same wavelength. Instead, in working at temperature T', one should use for comparison with a wavelength λ at temperature T a wavelength λ' given by

$$\lambda' T' = \lambda T. \tag{2.1}$$

Secondly, the spectral emittances for wavelengths corresponding in this fashion vary with temperature according to the relationship

$$e_\lambda/e_{\lambda'} = T^5/T'^5. \tag{2.2}$$

In particular, if e_λ (for fixed absolute temperature T) has a maximum value $e_{\lambda,\max}$ at some wavelength λ_m, then the values of $e_{\lambda,\max}$ at various temperatures satisfy the relation

$$e_{\lambda,\max}T^{-5} = \text{const.}, \tag{2.2a}$$

while λ_m satisfies the relation

$$\lambda_m T = \text{const.} \tag{2.1a}$$

Equations (2.1) and (2.2) are known as *Wien's displacement laws*. Together, although Wien did not note it, they imply that the expression for e_λ must be of the general form

$$e_\lambda = \lambda^{-5}f(\lambda T), \tag{2.3}$$

or, equivalently,

$$e_\lambda = T^5F(\lambda T),$$

where $F(\lambda T) = (\lambda T)^{-5}f(\lambda T)$. The specific form of the function $f(\lambda T)$ or $F(\lambda T)$ is not determined by these arguments; merely the fact that they depend on λ and T only through the product λT. It is worth noting that the Stefan-Boltzmann law can be derived from either of these expressions.

This was the most that could be established on the basis of classical theory without the addition of more detailed hypotheses. Wien, however, in a later paper (1896), went on to obtain an explicit form for e_λ by making explicit assumptions about the process of radiation from a molecule.[6] His result was

$$e_\lambda = C\lambda^{-5}e^{-c/\lambda T}, \tag{2.4}$$

consistent with Eq. (2.3) and in support of Paschen's empirical expression. The result was not universally accepted, however, as some physicists questioned the validity of the assumptions needed.

One person who did accept it, and who even provided additional arguments to support it, was Max Planck. Planck had been attracted to the problem of blackbody radiation by the "absolute" character of the distribution law—its independence of the material of the cavity walls. He made use of this independence in his work on the problem by taking as the walls an assemblage of harmonic oscillators. His approach was always through the thermodynamics of the assemblage, with particular emphasis on the thermodynamic quantity called *entropy*.[7] The precise definition of this quantity need not concern us here; the essential point is that it is a measure of the disorder in a system, and that the second law of thermodynamics, with which Planck was thoroughly familiar, states that the natural tendency of an isolated system is to change in such a way that its entropy increases. The implication is that the state of equilibrium is the state for which the entropy is a maximum; and it has already been mentioned that for a cavity, the equilibrium state is characterized by the cavity being filled with blackbody radiation. Planck's task, therefore, was to calculate the entropy of his assemblage.

In his initial work, published in 1899, Planck was not familiar with the "disorder" interpretation of entropy. He considered the entropy of an individual oscillator, and sought to relate the entropy to the energy U of the oscillator. He found that a basic quantity was the curvature R of the graph[8] of this relationship, and an erroneous assumption led him to conclude that

$$R = -\alpha/U, \tag{2.5}$$

where α is a positive quantity that might depend on frequency. The radiation law to which this led was just Eq. (2.4), with $\alpha = \lambda/c$.

One other theoretical attempt deserves mention, although it actually postdated the experimental work. In 1900, Lord Rayleigh approached the problem in the following way: A cavity can vibrate—i.e., sustain standing waves—only at fixed frequencies determined by its geometry, just as a string can vibrate only at fixed frequencies determined by its length. The radiation in a cavity must be composed of a superposition of the standing waves possible for the cavity. It is possible to calculate the number of different standing-wave modes with wavelength between λ and $\lambda + d\lambda$ that can exist per unit volume in the cavity. (The symbol $d\lambda$ here does not denote a true mathematical infinitesimal, but is used only to indicate that the range of wavelengths included is small compared with its end points.) According to a famous principle of classical statistical mechanics, the law of equipartition of energy, each of these modes should have the same average energy, namely an amount kT where T is the absolute temperature and k is a universal constant. These considerations lead to an expression for the energy per unit volume in the cavity per unit wavelength range, and thence to an equation for the spectral emittance,

$$e_\lambda = 8\pi ckT\lambda^{-4}. \tag{2.6}$$

This equation is consistent with Eq. (2.3). However, it has one serious defect: It yields an infinite value for the total energy in the cavity.[9] The infinite value is due to larger and larger contributions from shorter and shorter wavelengths. Since short wavelengths are associated with ultraviolet radiation, the divergence is often referred to as the "ultraviolet catastrophe." (In the light of modern knowledge, a more appropriate name might be the "gamma-ray catastrophe.")

In summary, the situation just before the turn of the century was this: There was a form for e_λ given by classical theory, presumably reliable but unspecific. There was a specific form obtained from additional hypotheses of uncertain validity, but supported by thermodynamic arguments. And there were several empirical approximations. It was natural that someone should try to improve on matters.

The definitive experimental work, carried out by Lummer and Pringsheim in Charlottenburg, Germany, used a method justified by a slight modification of the discussion that gave the equivalence between cavity radiation and blackbody radiation. Suppose that in one wall of a cavity such as envisioned in that discussion, there is a hole whose area is small compared with that of the walls. Any radiation that falls on the hole from outside is almost certain to pass into the cavity, there to be trapped by continual reflection back and forth by the walls (with some absorption at each reflection), so that the hole is in that respect a good approximation to a black body[10]—precisely the situation described in the third paragraph of this chapter. Moreover, the radiation coming out of the hole will be a representative sample of the radiation in the cavity, which has already been established as equivalent to black-body radiation; a study of its properties therefore gives the properties of the latter. The procedure appears to be simple in principle, but it required considerable care in its execution.

The first step, reported in 1897, was to verify the Stefan-Boltzmann law. Two cavities were used, a copper one for temperatures up to 877°K and an iron one for temperatures from 799°K to 1561°K. The copper cavity was immersed in a molten mixture of sodium nitrate and potassium nitrate; the temperature of the bath could be held constant to within one or two degrees for as long as half an hour by control of the supply of gas to the heating flame. The iron cavity was heated by means

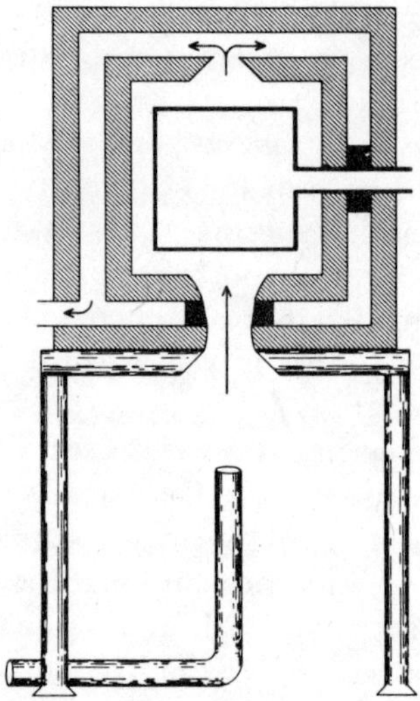

FIG. 2.1 Diagram of the double-walled oven as used for heating the iron cavity.

of a special double-walled oven shown in Fig. 2.1, in which the hot gases from the flame passed around the cavity inside the inner wall of the oven, then between the two walls, and then into the chimney flue. Temperatures up to 755°K were measured by means of mercury thermometers; higher temperatures, by a thermocouple.

The radiant power was measured by means of a bolometer. In this device, the radiation whose power is to be determined falls on one of two blackened platinum wires and is absorbed, raising the temperature of the wire and therefore its electrical resistance. The increase in resistance is measured by comparison with the resistance of the other wire.

Extensive precautions were taken to ensure that the energy recorded came only from the cavity, and to correct for possible variations in the fraction of radiation absorbed in the air along the path from cavity to sensing element. In fact, the only difficulty that was not almost entirely overcome was that of achieving truly uniform temperatures in the iron cavity. The final conclusion, based on observations over a range of temperature whose extremes differed by a factor of four, was that the Stefan-Boltzmann law is valid.

Lummer and Pringsheim then proceeded to a study of the more speculative question of how the spectral emissivity varies with wavelength at a given temperature. The results were reported in a series of three papers in 1899 and 1900. Again the basic concept was simple and the basic procedure straightforward, to analyze the radiation with a prism spectrometer before measuring the power, and only the several precautions and corrections were complicated. Various cavities were used, at temperatures from 85 to about 1800°K. The lower temperatures were achieved by immersion in liquid air (85°K), boiling water (373°K), and molten saltpeter (around 600°K, depending on exact composition). High temperatures, up to about 1800°K, were obtained by electrical heating.

At such temperatures, by far the largest fraction of the radiation is in the infrared region of the spectrum; the range of wavelengths studied was from about 1 micron to about 18 microns (visible light covers the range of wavelengths from about 0.4 microns for violet to about 0.7 microns for red). A substantial difficulty in this spectral region is that water vapor and carbon dioxide, both normally present in the atmosphere, absorb strongly near certain wavelengths, especially around 1.8, 2.7, and

4.5 microns. In the earliest work, Lummer and Pringsheim merely attempted to correct for this effect. Later, they enclosed the spectrometer and bolometer in a container in which the air was dried and chemically purged of carbon dioxide, so that the necessary correction was greatly reduced. As in the work on the Stefan-Boltzmann law, strict precautions were taken to ensure that only the radiation of interest fell on the bolometer.

One method of presenting the results is simply as a curve showing e_λ as a function of λ for various temperatures, as in Fig. 2.2 (from the second of the three articles). From such curves, Lummer and Pringsheim determined both the wavelength λ_m at which e_λ was maximum, and the maximum value $e_{\lambda,\max}$, for testing Eqs. (2.1a) and (2.2a). It was simply a matter of seeing whether the appropriate combinations of factors were indeed constants. Already in their first report, they felt justified in making the following statement:

> It can therefore be regarded as proven by this series of observations that for the radiating body employed, the maximum energy increases with the fifth power of the absolute temperature. Also the equation $\lambda_m T = A$ can be considered proven, since the deviations of A from the average value lie within the observational errors possible from the determination of λ_m.

Actually, it is not entirely clear that Lummer and Pringsheim regarded the displacement laws as in question, subject to proof or disproof. They later stated, "The fulfilment of these three laws" (the third being the Stefan-Boltzmann law) "is the *conditio sine qua non* if one wishes to draw from the radiation measurements any conclusion whatever about the form of the spectral equation (energy curve)." In other words, by this time at least they regarded the equations as more reliable than their measurements; in fact, they discarded one series of observations because the maximum value of e_λ varied as $T^{5.2}$ rather than as T^{5}.[11]

Wien's formula for e_λ, Eq. (2.4), though it was on somewhat less firm ground, was essentially the only explicit one available, and so Lummer and Pringsheim compared their results with it. Figure 2.2 shows, in addition to the set of experimental plots, the curves represented by Wien's formula for the values of T

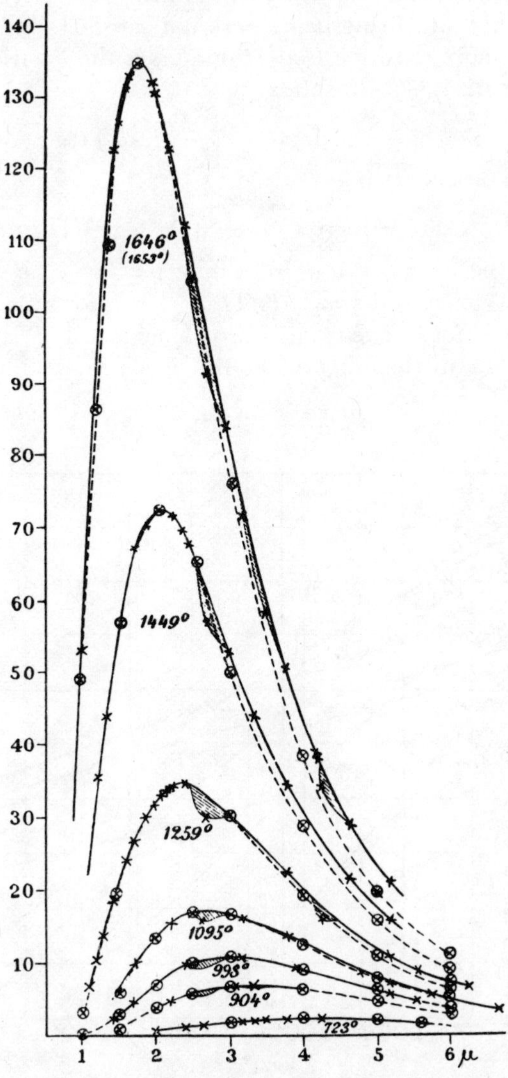

FIG. 2.2 Comparison of Lummer and Pringsheim's experimental data (plain crosses and solid curves) with Wien's formula (circled crosses and dashed curves). The shaded areas show the absorption by water vapor and carbon dioxide in the air. [*Verhandl. Deut. physik. Ges.* 1, p. 217 (1899), Fig. 1.]

used. The agreement between experiment and theory looks fair, but Lummer and Pringsheim were not satisfied, and devised the following more sensitive test: If one takes the logarithm of both sides of Eq. (2.4),[12] the result is

$$\log e_\lambda = \log(C\lambda^{-5}) - (c/\lambda T) \log \mathrm{e},$$

which can be rewritten as

$$\log e_\lambda = \log(C\lambda^{-5}) - [(c/\lambda) \log \mathrm{e}](1/T).$$

This has the form $y = a + bx$, where $y = \log e_\lambda$, $a = \log(C\lambda^{-5})$, $b = -(c/\lambda) \log \mathrm{e}$, and $x = 1/T$. Thus Wien's formula implies that when $\log e_\lambda$ for a fixed λ is plotted against $1/T$, the resulting curve, called an *isochromat,* should be a straight line; the slope

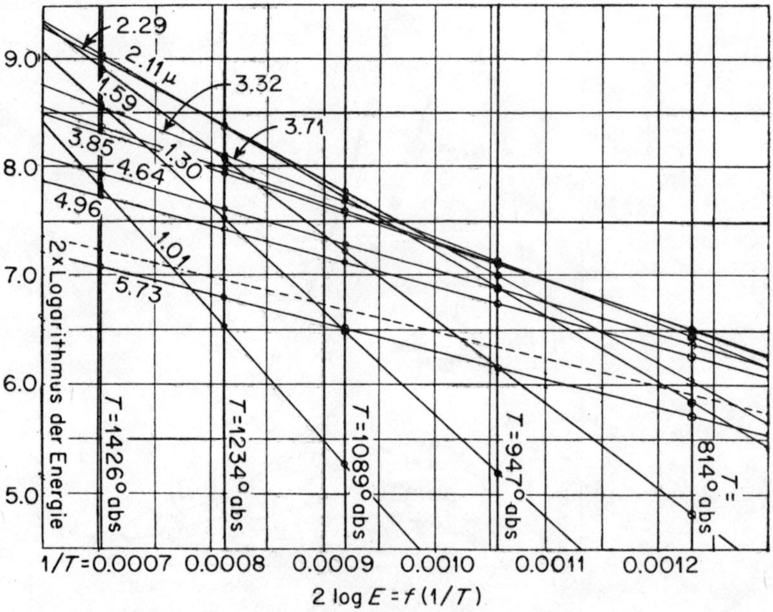

FIG. 2.3 A set of isochromats from Lummer and Pringsheim's first report on black-body radiation. [*Verhandl. Deut. physik. Ges.* 1, p. 38 (1899), Fig. 2; lettering replaced for legibility.]

of the line is proportional to *c*, and the intercept on the log e_λ axis can be used to compute *C*. The nature of the arguments used to deduce Wien's formula was such that the quantity *c* should be a constant of nature, while *C* might vary from one series of observations to another but should be constant throughout any one series.

In their earliest report, Lummer and Pringsheim presented isochromats that appeared to be straight lines, as shown in Fig. 2.3. The values of *C* and *c*, however, varied with wavelength. This was not as it should have been according to Wien's formula, but Lummer and Pringsheim were not sufficiently confident in their procedures to regard the test as conclusive. By the time of the second report, they were doubtful enough of the validity of Eq. (2.4) that they looked for, and reported finding, curvature in the isochromats (they did not show any isochromats in the second report). They say, however, "Nevertheless, before we pass judgment on the validity of the Wien-Planck equation, we consider it necessary to extend the studies over a larger temperature interval and wavelength range."[13] Finally, in the third report, the evidence had become unquestionable; the curvature in the isochromats was obvious (see Fig. 2.4). They firmly conclude that the Wien-Planck law is invalid, and say, "Likewise, with it, the foundation is pulled out from under all those extensive consequences that people have derived from the Wien-Planck equation."

Stimulated by these results, and by similar results obtained by his colleagues Rubens and Kurlbaum at Berlin, Planck attempted to modify his own theoretical work. Just how he arrived at his result, apart from the fact that it was by wholly empirical means, is not known; even Planck himself gave two different versions. In his first presentation, he told of constructing

> . . . completely arbitrary expressions for the entropy, which, although more complicated than the Wien expression, still appear to meet all the requirements of thermodynamic and electromagnetic theory as satisfactorily as that does.
>
> Among the expressions so drawn up, one struck me especially as being closest in simplicity to the Wien expression and . . . truly deserving of being tested next after that.

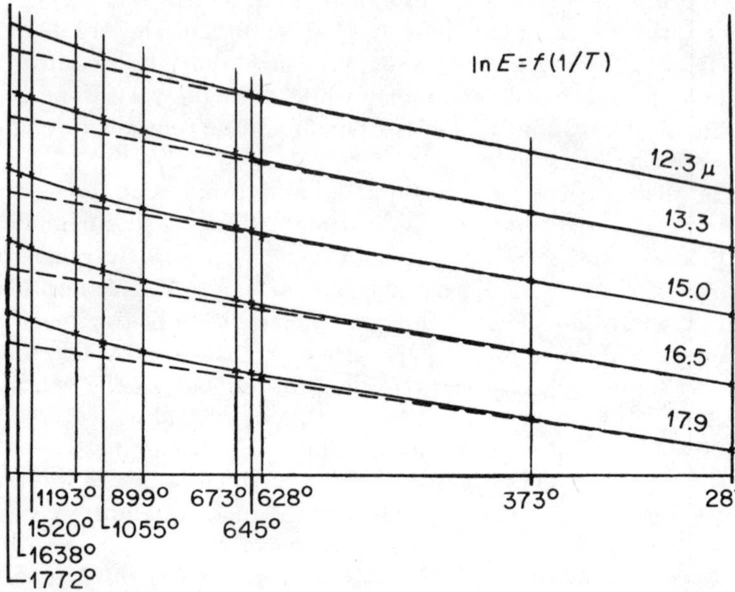

FIG. 2.4 A set of isochromats from Lummer and Pringsheim's third report. [*Verhandl. Deut. physik. Ges.* 2, p. 170 (1900), Fig. 1; lettering replaced for legibility.]

In other words, he had chosen a sort of next approximation after Eq. (2.5), such as might be obtained from an expansion of $1/R$ in powers of U. Later, however, he told [14] of having tried what he called "a luckily guessed interpolation formula" between Eq. (2.5), known to be valid for small U, and a variation of R with $1/U^2$ for large U. This latter form of variation was implied by Rubens and Kurlbaum's finding that for large values of the product λT, e_λ varied approximately linearly with T. By whatever process, Planck was led to replace Eq. (2.5) by

$$R = \alpha/U(\beta + U),$$

where α and β are constants. This gives a radiation law of the form

$$e_\lambda \propto \lambda^{-5}(e^{b/\lambda T} - 1)^{-1}, \tag{2.7}$$

where b is an unspecified constant. The equation proved to fit the experimental results very well. This result was presented before the Deutsche Physikalische Gesellschaft (German Physical Society) in October 1900.

Planck was able to justify Eq. (2.7) in December 1900, but only after he had come to a more complete understanding of entropy. Boltzmann had recognized as early as 1877 that the entropy of a state is a measure of the probability of the occurrence of that state. The probability, in turn, can be found simply by counting the number of different microscopic arrangements—in the present case, the number of ways of assigning energies to the individual oscillators—compatible with the given over-all state, and by assuming that each microscopic arrangement is equally probable. In order for the counting process to be possible in the case at hand, however, the energy cannot be a continuous variable but must be parceled out in multiples of a basic unit ϵ, so that $U = n\epsilon$. When Planck put these concepts into his treatment, he found that the entropy S depended on U and ϵ only through the combination U/ϵ. Wien's displacement laws, on the other hand, implied that $S = f(U/\nu)$, where ν is the frequency of the oscillator. Consequently, one must have $\epsilon = h\nu$: *The energy of an oscillator must be an integral multiple of a basic unit proportional to the frequency.* The resulting radiation law was

$$e_\lambda = \frac{8\pi c^2 h}{\lambda^5(e^{hc/k\lambda T} - 1)}, \tag{2.8}$$

where h is Planck's constant and k is another universal constant, and c is the speed of light.

This approach had several striking features. Not only does Eq. (2.8) fit the experimental results very well. In addition, when the wavelength or temperature is small so that the exponent of e is large,[15] it reduces to the Wien-Planck law, Eq. (2.5); while if λ or T is large, the equation reduces to Rayleigh's law, Eq. (2.6).[16] Moreover, the use of the entropy concept enabled Planck to deduce a value for what is now referred to as Avogadro's number and thus also for the charge on the electron; again, the values for these and other related quantities were in good agreement with the values obtained from other methods.

Oddly enough, however, it was not so much this agreement that convinced Planck. The second work was presented, like the first, before a meeting of the Deutsche Physikalische Gesellschaft; and Planck made these comments: "When I had the honor several weeks ago to direct your attention to a new formula . . . , I based my opinion of the usefulness of the formula . . . especially on the simple structure. . . ." He went on to say that he now intended to describe briefly how the formula could be deduced, but cautioned the audience that, "I do not place any value on the proof of necessity and of the easy practical applicability, but only on the clarity and unequivocal nature of the directions given for the solution of the problem."

The beginnings of quantum theory, then, lay in an experiment whose results could be understood only by the introduction of an idea foreign to classical theory: that in some systems, energy is not infinitely subdivisible, but is exchanged with the rest of the universe only in discrete amounts, or *quanta.*

FOOTNOTES

[1] This figure of speech is meant to imply only that the entire essential content of the theory was already present in Einstein's original articles. For a treatment of the special theory, see Robert Katz, *An Introduction to the Special Theory of Relativity,* MOMENTUM Book No. 9 (Van Nostrand, 1964).

[2] Actually, of course, we would not be able to see the object at all—still assuming that it was not luminous—but only a chunk of space from which no light reached us.

[3] It must be emphasized that this is *not* the same as reflection. The characteristics of reflected light are determined partly by the properties of the reflecting body but also partly by those of the incident light. Only in very special situations, however, are the characteristics of radiation affected by those of incident light.

[4] This fact is reflected in common language, where "red hot" and "white hot" denote different, though vague, temperatures.

[5] The base of natural logarithms, 2.71828 . . . , is denoted in this book by a roman e. This is not a universal practice, however, and the reader should not be disturbed by the use of a single symbol for several meanings. Another commonly used symbol for this number is the italic *e,* which has already been used to designate spectral emittance, and will be used later in the book as the symbol for the charge on the electron.

[6] In a sense, the use of the idea of molecules carried this work outside the realm of truly classical physics; cf. Chap. 1.

[7] For a further discussion of the concept of entropy, see Mark W. Ze-

mansky, *Temperatures Very Low and Very High,* MOMENTUM Book No. 6 (Van Nostrand, 1964), Chap. 2.

[8] The curvature at any point on a smooth curve is defined as the reciprocal of the radius of the circle that most closely fits the curve at that point.

[9] It is interesting to note that this was not the first indication of failure of the law of equipartition of energy. Maxwell had noticed as early as 1859 that the law was incapable of providing an adequate explanation of the ratio of the specific heats of a gas.

[10] The closeness of the approximation in this case depends on the fact that incident radiation will *fail* to be absorbed into the cavity only if one of its first few encounters with the walls is such as to reflect it out again. The likelihood of that happening is determined by the microscopic geometry of the walls and is independent of the wavelength of the radiation. In contrast, the closeness with which a real black object such as a lump of carbon approximates an ideal black body is determined by the probability that the incident radiation will be absorbed rather than reflected at its first and only encounter with the surface of the object; this is dependent on the wavelength.

[11] They state that this may have been the result of a poor adjustment which allowed the spectrometer to "look at" part of the cooler outer surface of the cavity as well as at the interior.

[12] The reader is reminded that the logarithm of a number y is the number a such that $10^a = y$; that therefore $10^{\log y} = y$; that if two numbers are equal, so are their logarithms; that $\log(xy) = \log x + \log y$; and that $\log(x^s) = \log[(10^{\log x})^s] = \log(10^{s \log x}) = s \log x$.

[13] At this stage, their measurements extended only from 620 to 1650°K, and from 1 to 6 microns in wavelength.

[14] In his Nobel lecture.

[15] In this case, unity can be neglected relative to the exponential.

[16] This can be seen by making use of the fact that $e^x \approx 1 + x$ when x is small.

BIBLIOGRAPHY

All but one of the papers treated in this chapter are in German, and only one seems to be available in translation. Consequently, only papers of special significance are listed here.

The concept of a black body is introduced, and Kirchhoff's law is proved, in G. R. Kirchhoff, *Annalen der Physik* **109,** 275–301 (1860) [translation, by F. Guthrie: *Philosophical Magazine* **20,** 1–21 (1860)].

Wien's displacement laws [Eqs. (2.1) and (2.2)] are deduced in W. Wien, *Annalen der Physik* **52,** 132–165 (1894).

Planck's derivation of Wien's distribution law, Eq. (2.4), is given in M. Planck, *Annalen der Physik* **1,** 69–122 (1900).

The experimental work with which this chapter is concerned is reported in O. Lummer and E. Pringsheim, *Verhandlungen der Deutsche Physikalische Gesellschaft* **1,** 23–41, 215–235 (1899); **2,** 163–180 (1900).

Planck's initial presentation of the correct distribution law is given in M. Planck, *Verhandlungen der Deutsche Physikalische Gesellschaft* **2,** 202–204 (1900); the derivation of it on thermodynamic grounds, in M. Planck, *ibid.* **2,** 237–245 (1900).

An extensive review of the history of these developments is given in Max Jammer, *The Conceptual Development of Quantum Mechanics* (McGraw-Hill, New York, 1966), Sec. 1.2.

See also *The World of the Atom,* Vol. 1, pp. 462–501.

3 Transmutation of Elements

The atomic concept as it was known at the end of the nineteenth century was almost identical in its over-all form, if not in detail, with what had been proposed nearly twenty-five centuries before by the Greek philosopher Leucippus. The universe consisted of empty space and matter; matter was composed of atoms. The number of individual atoms was extremely large—infinite, according to the Roman author Titus Lucretius Carus, whose book *De Rerum Natura,* or *On the Nature of Things,* is probably the outstanding early exposition of Leucippus's ideas. The number of different kinds of atoms, however, was finite, and presumably not very large. The properties of a material object were determined by those of the atoms of which it was composed.[1] The atoms themselves were indestructible and immutable.

The intervening centuries had had their effects, of course. The chemical concepts of element and compound had become well established, and it had become accepted within the atomic framework that the smallest unit of a compound was a molecule that would contain two or more atoms—that an aggregate of identical atoms formed an element. It was even recognized that some changes could be produced in atoms, electrons being attached or detached in the formation of ions; but such a change did not alter the basic nature of the atom that underwent it. An atom of sulfur did not become an ion of chlorine, but of sulfur.

The phenomenon of radioactivity, discovered in 1896 by Henri Becquerel, was to force drastic changes in this picture. For one thing, it was to provide the tool by which the atom itself would be found to have a structure and to consist mostly of empty space.

Even more startling, it was to be exposed as the adjunct of a change in the nature of atoms, while at the same time providing an extremely sensitive means of deducing that such a change had occurred. The discovery of this second aspect by Ernest Rutherford (later created Baron Rutherford of Nelson) and Frederick Soddy, which would lead to Nobel prizes for both men, will be described in this chapter.

It must be noted at the outset that this work was markedly different in one respect from the others described in this book: The researchers had no conception of what they were looking for, not even an inkling of what to expect. All the other cases represent, in essence, measurements of well-defined quantities, carried out against a background of previous experiment and theory which naturally suggested the measurement and which often indicated what the result should be. Here, in contrast, the investigators were groping, trying anything that looked as if it might provide information. The published reports are strongly flavored by this attitude. Unfortunately, to attempt to reproduce it here would unduly lengthen the book, and it has been necessary to be more restricted in quoting, and to do more abridging and summarizing, than is required in other chapters. The essential aspects, however, are retained.

The investigation was basically a study of the radioactive properties of the element thorium. Becquerel's original discovery had involved uranium; but subsequent work by the Curies had revealed the existence of a number of other radioactive elements associated mineralogically (and, it is now known, radiochemically) with uranium.[2] In particular, the radioactive nature of thorium was recognized about two years after Becquerel's discovery. Rutherford, then at the Cavendish Laboratory at Cambridge University, immediately took an interest in the new activity and its comparison with uranium; as an incidental point in a long article dealing with the radiations from uranium, written in 1898, he remarked that thorium appeared variable as a source of activity—in contrast to uranium, which was quite constant.

By the time that paper was published, in January, 1899, Rutherford had become MacDonald Professor of Physics at McGill University in Montreal. There he undertook a thorough study of the radiations from thorium. Soddy, a Demonstrator in Chemistry, was his chief collaborator in this research. The results were presented in a series of six papers: two in the *Philosophical Magazine,* in January and February, 1900; two in the *Transactions of the Chemical Society,* in 1902; and two more in the *Philosophical Magazine,* also in 1902, the last two being strongly repetitive of the preceding two. Soddy was coauthor of the last four.

The investigation started with the curious variability of the radiation. The first of the six papers reports the following observation: "The intensity of thorium radiation, when examined by means of the electrical discharge produced, is found to be very variable; and this inconstancy is due to slow currents of air produced in an open room. When the apparatus is placed in a closed vessel, to do away with air currents, the intensity is found to be practically constant." It should be mentioned here that the method of comparing intensities, which had been used in earlier work and described in an earlier paper, was based on the fact that the radioactive rays produced ionization in the air through which they passed. What was actually measured was the rate at which the ionization caused the discharge of what amounted to a parallel-plate air capacitor. The plates were horizontal, one about 4 centimeters above the other; a layer of the radioactive material was placed on the lower one. The plates were initially given a difference of potential of about 100 volts; and the rate at which the difference decreased, measured by means of a sensitive electrometer, was regarded as proportional to the "intensity" of the radiation.[3]

The next step was to study the absorption of the radiation. A homogeneous radiation, passing through a medium that is not completely opaque to it, can be expected to undergo absorption, with the fraction that has been absorbed increasing exponentially with the thickness of the medium that has been traversed. The rate of increase depends partly on the absorber but also on the radiation, and this method was already in common use to help characterize radioactive radiations. In the case of thorium, however, with ordinary foolscap paper as absorber, the results were peculiar in that they depended on whether the radiation came from a "thin" layer—a sprinkling of fine powder—or a "thick" layer. In the former case, the absorption followed the expected exponential law fairly well, with the intensity reduced to 0.08 times its initial value by a thickness of 0.0081 centimeter of paper. For a thick layer of radioactive material, however, 0.008 centimeter of paper reduced the intensity only by a factor of 0.74, and additional paper had little effect.

Layers of different thicknesses also displayed a difference in sensitivity to air currents. "On directing a slight current of air between the test plates, the rate of discharge due to a thick layer of thorium oxide[4] is greatly diminished. . . . Under similar conditions with uranium, the rate of leak is not significantly affected." This last statement established that the effect is not the result of the method and conditions of measurement, but is really connected with the behavior of thorium. "With a thin layer of oxide, the diminution of the rate of leak is small; but with a thick layer of oxide, the rate of leak may be reduced to less than one-third of its previous value."

There follows a sentence, still referring to the thick layer, which is especially significant: "If two thicknesses of foolscap paper are placed over the thorium oxide, the resulting rate of leak between the plates may be diminished to less than $\frac{1}{20}$ of its value by a slight continuous blast of air." It is to be recalled that paper alone had little effect on the intensity of radiation from a thick layer of oxide; but the additional action of a stream of air made the radiations from thick and thin layers similar in their absorption by paper.

In summary, it appeared that the radiations from a thin layer of oxide were normal, obeying the usual absorption law and not affected by air currents. Those from a thick layer, on the other hand, while they apparently included a small component of the normal variety, were mostly of another variety, not absorbed in the normal fashion and strongly affected by air currents. Rutherford was led to the following deduction:

> The phenomena exhibited by thorium compounds receive a complete explanation if we suppose that, in addition to the ordinary radiation, a large number of radioactive particles are given out from the mass of the active substance. This 'emanation' can pass through con-

siderable thickness of paper. The radioactive particles emitted by the thorium compounds gradually diffuse through the gas in its neighbourhood and become centres of ionization throughout the gas. The fact that the effect of air currents is only observed to a slight extent with thin layers of thorium oxide is due to the preponderance, in that case, of the rate of leak due to the ordinary radiation over that due to the emanation. With a thick layer of thorium oxide, the rate of leak due to the ordinary radiation is practically that due to a thin surface layer, as the radiation can only penetrate a short distance through the salt. On the other hand, the 'emanation' is able to diffuse from a distance of several millimetres below the surface of the compound, and the rate of leak due to it becomes much greater than that due to the radiation alone.

The explanation of the action of slight currents of air is clear on the 'emanation' theory. . . . Extremely minute motions of air, if continuous, remove many of the radioactive centres from between the plates. . . . The emanation continues to ionize the gas in its neighborhood for several minutes, so that the removal of the particles from between the plates diminishes the rate of discharge between the plates.

In summary, the thorium was giving off some substance which was itself able to produce ionization and which, if not a gas, was at least in a form that could be transported by a stream of air.

The remainder of the first paper was concerned primarily with the effect of various conditions on the production of the emanation. Three additional properties of the emanation itself were deduced, however. One was that its own radioactivity decayed exponentially with time, decreasing by a factor of $\frac{1}{2}$ approximately every minute. This was one of the first demonstrations of the decay of radioactivity; and it is quite possible that when Rutherford submitted the paper, in September of 1899, he had not learned of any other such work.[5] A second property, which Rutherford quite naturally describes as "very remarkable," was that "the positive ion produced in a gas by the emanation possesses the power of producing radioactivity in all substances on which it falls."

The third property was that the emanation was, or appeared to be, truly a gas. "The emanation passes through a plug of cotton-wool without any loss of its radioactive powers. It is also unaffected by bubbling through hot or cold water, weak or strong sulphuric acid. In this respect it acts like an ordinary gas.

An ion, on the other hand, is not able to pass through a plug of cotton-wool, or to bubble through water, without losing its charge." Evidently Rutherford had entertained the thought that the radioactivity might be the manifestation of some property possessed by an ion produced by the thorium radiation.

The second paper of the series was devoted to the "excited radioactivity," the activity referred to in the second property mentioned above. The essential results were that this excited activity was, "in some way, due to the 'emanation', or to something that accompanies it, but is not caused by the direct action of a radiation from thorium oxide"; that it also decayed, with a half-life of about eleven hours; and that it involved the deposition of some material substance, though in extremely minute quantities.

It was in the third paper of the series, in which for the first time Soddy was a collaborator, that the really crucial step was taken. The investigators posed themselves a series of questions. Most of these had to do with the production of the emanation; but one was a more complete determination of the nature of the emanation itself. Specifically, they wished "to see whether the emanation possesses chemical properties which would identify it with any known kind of matter." Here Soddy's training as a chemist and his knowledge of contemporary developments in chemistry were indispensable.

Presumably, the apparatus used for this part of the investigation was essentially the same as for the part dealing with production of the emanation.

[Figure 3.1] shows the experimental arrangement for comparing the emanating power of substances. These are placed in the form of fine powder in a shallow lead vessel inside the glass cylinder, *C,* 17 cm. in length and 3.25 cm. in diameter, provided with indiarubber corks. A current of air from a large gas-bag, after passing through a tube containing cotton wool to remove dust particles, bubbled through sulphuric acid in the vessel, *A*. It then passed through a bulb containing tightly packed cotton wool to prevent any spray being carried over. The emanation mixed with air was carried from the vessel *C* through a plug of cotton wool, *D,* which completely removed all the charged carriers carried with the emanation. The latter then passed into a long, brass cylinder 75 cm. in length and 6 cm. in diameter. The cylinder insulated on paraffin blocks was connected to one pole of a battery of small lead accumulators,[6] the other pole of which was connected to earth. Three electrodes *E, F, H,* of equal length were placed along the axis of the cylinder, supported by brass rods passing through ebonite corks in the side of the cylinder. The current through the gas, due to the presence of the emanation, was measured by means of a Kelvin quadrant electrometer of the White pattern.[7] . . . An insulating key was arranged so that either of the electrodes *E, F, H,* or all of them together, could be rapidly connected to one pair of quadrants of the electrometer, the other two being always connected to earth.

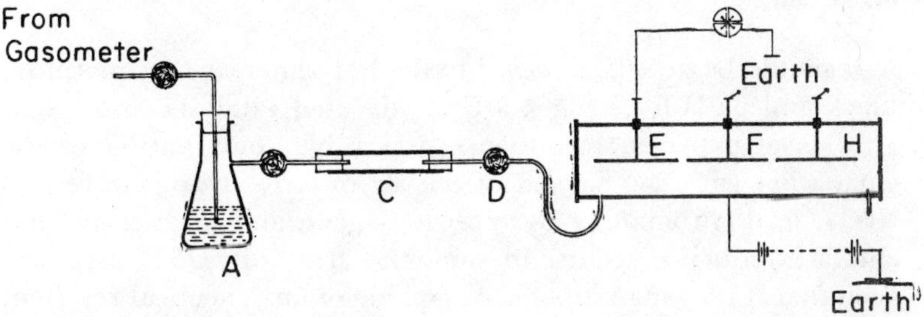

FIG. 3.1 Rutherford and Soddy's apparatus for studying the emanation from thorium. [*Phil. Mag. 4,* p. 569 (1902), Fig. 1; lettering replaced for legibility.]

With such an arrangement, using a battery of 50 volts potential, typical currents were of the order of a few times 10^{-11} ampere, giving scale deflections of 100 divisions (size unspecified) in a time of the order of 10 seconds. The current was proportional to the amount of ionization produced in the brass cylinder, which in turn was proportional to the amount of emanation present. The investigation of the properties of the emanation proceeded simply by seeing what kind of reaction it could be made to undergo.

The effect of temperature was first tried. The air containing the emanation, obtained in the usual way by passage over thoria, was led through the platinum tube[8] heated electrically to the highest attainable temperature, and also through the tube cooled by solid carbon dioxide and ether. The tube was then filled with platinum black, and the emanation passed through in the cold, and with gradually increasing temperatures, until the limit was reached. The effect of the intense heat was to convert the platinum black completely into platinum sponge.

In another experiment, the emanation was passed through a layer of red hot lead chromate in a glass tube. The current of air was replaced by a current of hydrogen and the emanation sent through red hot magnesium powder, and red hot palladium black, and, by using a current of carbon dioxide, through red hot zinc dust. In every case, the emanation passed without sensible change in the amount. If anything, a slight increase occurred, owing to the time taken for the gas current to pass through the tubes when hot being slightly less than when cold, the decay *en route* being consequently less. It will be noticed that the only known gases capable of passing in unchanged amount through all the reagents employed are the recently discovered gases of the argon family.

It should be noted, in regard to the last sentence, that platinum black and palladium black are widely used catalysts, and might have been expected to enhance a possible combination of the emanation with the oxygen of the air or with hydrogen, respectively; lead chromate is an oxidizing agent, and magnesium and zinc are reducing agents. In sum, the array of agents used was such that if the emanation were capable of any chemical reaction, it should have undergone at least one in the course of these attempts.

Soddy, many years later, described the experiment and his and Rutherford's reactions in this way:[9]

I simply inserted into the air-stream, between the thorium compound and the ionization chamber in succession, a number of powerful chemical reagents such as magnesium powder, platinum black, lead chromate, and zinc dust at temperatures up to white heat—one or other of which would have absorbed every known gas before the discovery of argon. The emanation passed through all of them completely unabsorbed and unaffected.

Eight years earlier this inability to absorb the gas would have conveyed nothing, even to a chemist, but in 1900 I knew that it meant that thorium was changing spontaneously into an argon gas.

. . . My mind was always occupied with transmutation. That is natural; I was a chemist. . . . Also at that time I had been working on the lectures on gas analysis which I had been asked to give in the university. That was why, perhaps, when Rutherford showed me the emanation which was not thorium, nor alpha nor beta particles, but which could be blown about, I drew his attention to the fact that it would be a gas. . . .

I was, of course, tremendously elated to have discovered transmutation—the goal of every chemist of every age . . . yet when the time came to investigate the phenomenon the whole thing seemed too devastatingly simple. The fact that this was in reality transmutation flashed through my brain and I could hardly believe what I knew to be true.

. . . after interpreting the meaning of the discovery which was that the atom was quite definitely disintegrating spontaneously, I was overwhelmed with something greater than joy—I cannot very well express it—a kind of exaltation, intermingled with a certain feeling of pride that I had been chosen from all chemists of all ages to discover natural transmutation.

I remember quite well standing there transfixed as though stunned by the colossal import of the thing and blurting out—or so it seemed at the time:

"Rutherford, this is transmutation: the thorium is disintegrating and transmuting itself into an argon gas."

The words seemed to flash through me as if from some outside source.

Rutherford shouted to me, in his breezy manner, "For Mike's sake, Soddy, don't call it *transmutation*. They'll have our heads off as alchemists. You know how they are."

After which he went waltzing round the laboratory, his huge voice booming "Onward Christian so-ho-hojers," which, as H. R. Robinson[10] declared, was more recognizable by the words than by the tune.

. . . I only want to show you how our brains were working, mine on transmutation and gases, Rutherford's on thorium and alpha ray emissions.

I realized that if the thorium emanation was transmuting itself[11] into another element—an argon gas—no doubt other elements were undergoing a process of natural transmutation in much the same way.

This was the start of the disintegration theory of radioactive substances.

The investigators were not yet ready to announce their findings in such blunt terms. For one thing, although they had carried out a test designed to eliminate the possibility, they still considered that "it may be that one of the inert constituents of the atmosphere is rendered radioactive in the presence of thoria and so constitutes the emanation." More to the point, probably, was the fact that other experiments, directed at the question of what might affect the emanating power of various thorium compounds, had given extremely puzzling results.

In the first experiment, which we have so far not succeeded in repeating, by an accident in the conditions apparently, two fractions were separated from thorium which differed in their emanating power in the ratio of 200 to 1. The active fraction diminished to nearly a third of its original value in fourteen days spontaneously, whilst the activity of the inactive fraction was, to a large extent, regenerated by solution and reprecipitation.

. . . Attempts to repeat this result have so far led to the production of two more or less completely de-emanated [12] fractions, which, however, spontaneously increase in activity with time. . . .

These experiments involved preparing solutions of thorium nitrate and precipitating the thorium, either as thorium carbonate, thorium hydroxide, or a combination; the precipitate was removed from the liquid by filtering.

The production of preparations of such low emanating power led naturally to an examination being made of the filtrates and washings for radioactivity. It was found that these possess, when concentrated, both emanating power and radioactivity in considerable amounts, although from the nature of their production they should be chemically free from thorium. The behavior is quite general. . . . It does not matter [how the precipitation of the thorium was carried out]—the thorium-free filtrate invariably possessed emanating power, and when evaporated to dryness exhibited straight-line radioactivity[13] also in amounts very much greater than possessed by the same weight of thoria.

The results of a careful chemical investigation of the active filtrates . . . was to show that these contained no thorium, or at most only a minute trace, but another substance in very appreciable quantities. . . . It has not yet been obtained in sufficiently large quantities for an exhaustive chemical examination, and it is impossible at present to say what it may prove to be.

The new substance was tentatively dubbed "thorium-X," and a series of experiments was begun to attempt to learn more about it. By this time, Rutherford and Soddy had come to recognize that emanating power was a poor measure of the presence of the new substance, being subject to considerable alteration, and had started to rely much more heavily on ordinary radioactivity. "It has been shown that it is difficult to follow, by means of the value of the emanating power, the progress of the removal of the active material. When this was realized, attention was directed to the straight line radioactivity, which is generally unaffected by these changes of conditions and previous history which produce such profound alteration in the former property."

Starting, then, with seventy grams of thorium nitrate, they extracted from it about 0.023 gram of a salt of thorium-X which was as intensely active as 200 times as much ordinary thoria. The "true" thorium that was left was reduced in activity to only about a third of what it had been.

At this point, according to the next paper in the series, the Christmas holidays intervened, and these preparations were allowed to stand. When the investigators returned "about three weeks later it was found that the thorium hydroxide, which originally possessed only about 36 percent of its normal activity, had almost completely recovered the usual value. The active residues, on the other hand, . . . had almost completely lost their original activity." A more detailed study showed that the activity of thorium-X decays exponentially with a half-life of about 4 days, while the recovery of the hydroxide followed an inverted exponential curve. As Rutherford had shown in another connection, these shapes of the curves will be explained if two processes are supposed to be taking place:

"(1) That the active constituent ThX is being produced at a constant rate,

"(2) That the activity of the ThX decays geometrically with time."

The second of these points had already been established, and it remained to confirm the first. This was done in a straightforward way.

> If a period of 24 hours is allowed to elapse between the successive precipitations, the activity of the ThX formed during that time corresponds to about $\frac{1}{6}$ of the maximum activity of the total thorium employed. In three hours, the activity of the amount produced is about $\frac{1}{30}$. The rate of production of ThX worked out from these figures well agrees with the form of the curve obtained for the recovery of activity of thorium if the latter is taken to express the continous production of ThX at a constant rate and diminution of the activity of the product in geometrical progression with time.

Rutherford and Soddy already realized that it was thorium-X and not thorium itself that produced emanation. The emanation, in turn, was known to produce "excited" radioactivity. "This process must be going on in the matter of the thorium compound itself, and . . . contributes an important quota to the total radioactivity of the compound." The summary of the experimental situation was this:

> The constant radioactivity of thorium is maintained by the production of [ThX] at a constant rate. . . . The ThX is undergoing a further change, and one of the products is gaseous and in the radioactive state constitutes the emanation produced by the thorium compounds. The ThX further possesses the property of exciting radioactivity on surrounding inactive matter . . . [probably] the same as the excited radioactivity produced by the thorium emanation, which has been shown to be produced by ThX.

The crucial statements, however, were made in the last two papers of the series. In the first of these, the authors say, "Since . . . radioactivity is at once an atomic phenomenon[14] and accompanied by chemical changes in which new types of matter are being produced, these changes must be occurring within the atom, and the radioactive elements must be undergoing spontaneous transformation." More bluntly stated, their studies, groping and stumbling as they had sometimes been, had convinced them that radioactivity inevitably involves transmutation.

The final paper contains a suggestion which, while strictly subordinate to the topic of this chapter, was even more far-sighted:

> Enough has been brought forward to make it clear that in the radioactivity of thorium . . . we are witnessing the effect of a most complex series of changes, each of which is accompanied by the continuous production of a special kind of active matter. The complexity of the phenomenon gives rise to an important question concerning the fundamental relation between the changes which occur and radioactivity. So far it has been assumed, as the simplest explanation, that the radioactivity is *preceded* by chemical change . . . A slightly different view is at least open to consideration, and is in some ways preferable. Radioactivity may be an *accompaniment* of the change, the amount of the former at any instant being proportional to the amount of the latter. On this view the non-separable radioactivities of thorium and uranium would be caused by the primary change in which ThX and UrX are produced. The activity of ThX would be caused by the secondary change producing the emanation, the activity of the emanation by a tertiary change in which the matter causing the excited activity is pro-

duced, the activity of the latter being derived from still further changes. The law of the decay of the activity with time . . . in all cases but the primary then appears as the expression of the simple law of chemical change, in which one substance only alters at a rate proportional to the amount remaining. In the primary change the amount remaining is infinitely great compared with the amount that alters in a short time, and therefore the velocity of reaction is constant.

They admit that "further work . . . must be awaited before the connexion . . . can be considered exactly determined." But they had found the path that would lead Soddy, ten years later, to be one of three men[15] independently deducing the laws of radioactive change—natural transmutation—as they are accepted today.

FOOTNOTES

[1] This is not to say that the properties of megascopic matter *were* those of the component atoms, nor even that atoms necessarily had the same kinds of properties as bulk matter, but only that one determined the other. A sour taste, for example, might be the manifestation of needle-shaped atoms.

[2] The story of this work, as well as other aspects of the early history of radioactivity, is sketched by Wilfred B. Mann and S. B. Garfinkel, *Radioactivity and Its Measurement,* MOMENTUM Book No. 10 (Van Nostrand, 1966), Chaps. 1 and 2.

[3] It is now known that the rate of discharge depends on two properties, not one: the number of radioactive disintegrations per unit time, and the energy carried by each emitted particle. It was not until after the investigations being described, however, that the radiations were unmistakably shown to consist at least partly of charged particles; and it was even later, of course, that the quantitative aspects of their interactions with the air were established. It can be seen that the electrical method was quite reliable for comparison of different samples of the same material, but could give quite misleading results for comparison of different sources.

[4] The thorium was studied in the form of a compound, usually thorium oxide (thoria), presumably chosen as being convenient to prepare and easy to work with. In the descriptions, the terms "thorium," "oxide," and "thoria" are used almost interchangeably.

[5] The first recognition that radioactivity decayed with time was achieved by Elster and Geitel in 1899. Most of the radioactive substances that had been isolated and identified at that time decayed so slowly that the change would not ordinarily be noticed.

[6] This is identical in chemical arrangement and electrical action, though not in physical structure, to several automobile storage batteries connected in series.

[7] The operation of this instrument is described in Appendix A.

[8] This is the first mention of "the platinum tube," which must have been placed in the flow train between the cotton-wool plug *D* and the brass cylinder.

[9] *Pioneer Research on the Atom: The Life Story of Frederick Soddy,* by Muriel Howorth (New World Publications, London, 1958), pp. 82–84.

[10] A student, and subsequently a colleague, of Rutherford's a few years after the time of this work.

[11] *Sic;* compare the later quotation from the last paper of the series.

[12] That is, deprived of the ability to produce emanation.

[13] The "normal" kind, that was not carried away by the air stream.

[14] This, as they noted, was generally accepted by all workers in the field.

[15] The other two were K. Fajans and A. S. Russell.

BIBLIOGRAPHY

Ernest Rutherford, *Philosophical Magazine* **49,** 1–14 and 161–192 (1900); E. Rutherford and Frederick Soddy, *Transactions of the Chemical Society* **81,** 321–350 and 837–860 (1902), and *Philosophical Magazine* **4,** 370–396 and 569–585 (1902). The last two papers are highly repetitive of the preceding two, but omit some of the material. All six papers are reprinted in Sir James Chadwick, ed., *The Collected Papers of Lord Rutherford of Nelson* (George Allen and Unwin Ltd., London, 1962), Vol. 1, pp. 220–231, 232–259, 376–402, 435–456, 472–494, and 495–508, respectively; the fifth in the series is excerpted in *The World of the Atom,* Vol. 1, pp. 451–461.

4 *The Existence of Atoms*

Classical physics had no particular need for the concept of atoms. The development of the concept was, of course, followed and understood by physicists; in some areas of physics it provided at least a convenient terminology, and there were those who accepted it wholeheartedly and used it. But at the end of the nineteenth century there was no apparent compelling need to believe in atoms; and, as mentioned in Chapter 1, there were a number of scientists who did not—including some of the foremost chemists of the time, despite the fact that much of the evidence in favor of the atomic hypothesis was chemical.

The skeptics had their doubts cleared away in 1908 and 1909 by a single piece of research, carried out by Jean Perrin. His work, reported in barest outline in a series of papers in the journal *Comptes Rendus des Séances Hebdomadaires de l'Académie des Sciences* (Proceedings of the Weekly Meetings of the Academy of Sciences), and in detail in the *Annales de Chimie et de Physique,*[1] showed conclusively that Brownian motion implied the existence of molecules; and it included two distinct methods of determining Avogadro's number, the number of molecules in one gram mole, from the properties and behavior of emulsions. In recognition of this work, which is described in the present chapter, Perrin received the Nobel prize for physics in 1926.

Brownian motion is the name applied to a phenomenon discovered in 1827 by a botanist named Brown: that very small bits of solid matter (such as pollen grains), when suspended in a fluid, are in continual random motion. Perrin describes it this way: ". . . every particle situated in the liquid, instead of taking on, according to its density, a regular movement of falling or rising, is rather found animated by a completely irregular movement. It comes and goes, it stops, it starts again, *it rises,* it sinks, *it rises again,* without at all tending toward immobility." However, according to Perrin, "The odd phenomenon discovered by Brown did not attract much attention. Rather it was ignored for a long time by most physicists, and it can be supposed that those who had heard about it believed that it was analogous to the motion of dust motes that we see dancing in a sunbeam under the action of weak air currents caused by slight differences of temperature or pressure."

It is significant that those who did study the phenomenon, almost without exception, were drawn to the conclusion that it was molecular in origin. Perrin quotes the explanation given by one Father Carbonnelle in the late 1870's:

> In the case of a surface of some extent, the collisions of the liquid molecules, the cause of pressure, do not produce an agitation of the suspended body, because their aggregate effect acts equally on the body in all directions. But if the surface is smaller than the extent that can provide compensation of the irregularities, there is no longer reason to consider the mean pressure, one must take account of pressures that are unequal and varying continually from point to point, which the law of large numbers no longer reduces to uniformity, and whose resultant is not zero, but changes constantly in magnitude and direction.

It is this changing force that gives rise to the erratic motion of the particle. "Moreover," Father Carbonnelle continued, "the inequalities become more and more apparent in proportion as the body is imagined to be smaller, and consequently the oscillations at the same time become more and more lively. . . ." This aspect of the phenomenon will be discussed in more detail later in the chapter.

As has already been mentioned, the atomic hypothesis was well known in 1908, and had been for several decades. The quotation in the preceding paragraph shows how earlier workers on the phenomenon of Brownian motion had utilized the concept of molecular motion to account for the phenomenon. Perrin felt that the argument could be made equally well, and perhaps even more convincingly, the other way around—that the existence of Brownian motion could be used to deduce the existence of molecules. His line of reasoning was as follows:

What is really strange and *new* in Brownian motion is just that it never stops. At first this seems in contradiction to our daily experience with friction. If, for example, we pour a bucket of water into a tub, it will seem proper to us that after a little while the motion first possessed by the liquid mass has disappeared. However, let us analyze how this apparent equilibrium is reached: All the particles of the water at first had their velocities approximately equal and parallel; this arrangement is broken up as soon as some of the particles, striking the sides of the tub, bounce off in different directions with changed velocities, to be immediately deflected anew by their collisions against other parts of the liquid. Thus, some instants after the fall, all the parts of the water are still in motion, but it is already necessary to consider a rather small part in order for the velocities of its different points to have approximately the same direction and the same magnitude. This is easily seen by mixing into the liquid some grains of colored powder that take on more and more irregular motions relative to each other.

What we observe, therefore, as long as we can distinguish anything, is not that the movement stops, but that it becomes more and more disorderly, that it is distributed in a more and more irregular manner among smaller and smaller parts.

Does this disordering proceed indefinitely?

To have some information for judging this, in any event to follow this disordering as far as possible, we must give up observing with the naked eye and make use of a microscope, and for indicator grains, make use of microscopic granules. Now, these are precisely the conditions under which we perceive Brownian motion; we are, then, *assured* that the coherence of movement, so evident on the ordinary scale of our observations, does not proceed indefinitely, and, on the microscopic scale, we *prove* an equilibrium between coherence and decoherence. . . . And there appears to be no way to avoid the following conclusion:

Since the dispersal of motion in a fluid does not proceed indefinitely and is limited by a spontaneous reordering, the fact is that fluids are themselves composed of grains, *of molecules,* which can indeed take on, relative to one another, all possible motions, but to the interior of which no transfer of motion is possible. If such molecules did not exist, there would appear to be no way to have a limit on the disordering of motion.

These arguments seem impressive; but as always in physics, they were of no value unless they could be supported by evidence of a quantitative sort. Perrin now addressed himself to the problem of providing such evidence.

One of the consequences of the molecular theory is the law relating the average kinetic energy W of a gas molecule to the pressure p and volume V of its container:

$$W = \tfrac{3}{2}pV/n, \tag{4.1}$$

where n is the number of molecules present in the container. Now imagine a dilute solution separated from a body of pure solvent by a semipermeable membrane, one which offers no impedance to the passage of solvent molecules but which completely stops molecules of the solute. In such an arrangement, there is a pressure exerted on the membrane, known as osmotic pressure. Exactly the same line of reasoning that led to Eq. (4.1) can be applied to the molecules of the solute, with the result that an equation of exactly the same form is found to relate the mean kinetic energy of the solute molecules to the osmotic pressure. Moreover, the Dutch chemist van't Hoff had found that experimental values of osmotic pressure were just the same as the pressures that would have been exerted by the respective solutes if they had been gases of unchanged densities.[2]

In the case of gases, there is another relation between pressure and volume: the gas law,

$$pV = NRT,$$

where T is the absolute temperature, N is the number of moles of gas, and R is a universal constant. This equation together with Eq. (4.1) imply

$$W = \tfrac{3}{2}(NR/n)T. \tag{4.2}$$

The ratio of the number n of molecules to the number N of moles is also a constant, Avogadro's number, denoted by N_0. Equation (4.2) is thus a statement that the mean kinetic energy of a gas molecule depends only on the absolute temperature of the gas. According to the preceding paragraph, an exactly parallel statement is true for the molecules of solvent in a dilute solution.

This last statement is completely free of all reference to the nature of either the solvent or the solute. If one considers a solute which is itself a liquid, ethyl alcohol for example, then, according to Perrin, ". . . this indifference to the nature of the molecules of the liquid in which it is moving makes it almost impossible to believe that it will not have the same energy if it is in ethyl alcohol, that is, if it is one of the molecules of pure ethyl alcohol."[3] Much more significant to the present discussion is that there is no upper limit placed on the size of the solute molecules, and there seems to be no obstacle to extending the law to "a particle yet a little larger, itself formed of several molecules, in one word[4] a *grain of powder*." Apparently, then, Eq. (4.2), or more properly its equivalent form

$$W = \tfrac{3}{2}(R/N_0)T, \tag{4.3}$$

applies also to particles observed in Brownian motion.

Let us return to Perrin's own description:

Suppose that it were possible to realize an emulsion of completely identical grains, an emulsion of which I will say, for brevity, that it is *uniform*. . . .

Imagine a uniform emulsion in equilibrium, which fills a vertical cylinder of cross section s. The state of the horizontal slab bounded by the heights h and $h + dh$ would not be changed if it were entrapped between two pistons permeable to the molecules of water, but impermeable to the grains. . . . Each of these semipermeable pistons is subjected, by the impacts of the grains that it stops, to an osmotic pressure. If the emulsion is dilute, this pressure can be calculated by the same reasoning as for a gas or an extended solution, with the result that, if at height h there are n grains per unit volume, the osmotic pressure P will be equal to $\tfrac{2}{3}nW$, if W denotes the mean energy of the grains; it will be $\tfrac{2}{3}(n + dn)W$ at height $h + dh$. Now, the slab of grains under consideration does not sink; for this it is requisite that there be equilibrium between the difference of osmotic pressures, which acts upwards, and the total weight of the grains, diminished by the buoyant force that they experience, which acts downwards. Thus, denoting by φ the volume of each grain, Δ its density, and δ that of the suspending liquid, we see that[5]

$$-\tfrac{2}{3}s\,W dn = ns\,dh\,\varphi(\Delta - \delta)g$$

or

$$-\tfrac{2}{3}W\frac{dn}{n} = \varphi(\Delta - \delta)g\,dh,$$

which, by an obvious integration, gives rise[6] to the following relation between the concentrations n_0 and n at two points whose difference in heights is h:

$$\tfrac{2}{3}W \ln \frac{n_0}{n} = \varphi(\Delta - \delta)gh, \tag{4.4}$$

a relation that can be called the *equation of distribution* of the emulsion.

The point of the foregoing derivation is this: Equation (4.3) expresses the mean energy of a suspended grain in terms of the absolute temperature and a quantity characteristic of the atomic-molecular hypothesis, Avogadro's number. But the mean energy cannot be measured directly; the motion of the particles is too fast and too short-ranged to be followed. It is necessary to relate the mean energy to quantities that could be measured. Equation (4.4) expresses such a relation, obtained *without* reference to the atomic hypothesis. Perrin has this to say:

> In short, by whatever path it be, we find ourselves led to think that the mean energy of translation of a molecule is equal to that possessed by the grains of an emulsion. If then we find a method of calculating this granular energy in terms of measurable quantities, we shall have in the same stroke a method of judging our theory. Once the experiments are done, two cases can in fact present themselves: Either the numbers obtained will be substantially different from those that are given by the kinetic arguments summarized above, and, in this case, especially if the numbers change with the grains studied, the credibility of the kinetic theories will be reduced, and the origin of the Brownian motion will remain to be found; or the numbers will be of the order of magnitude predicted, and, in this case, not only do we have the right to regard the molecular theory of this motion as established, but also we can seek in our experiments a method, perhaps precise this time, of knowing molecular magnitudes.

The stage was thus set for the experiment. The concept, once again, was simple: to determine the concentrations at two levels in an emulsion of particles of known size and density and at known temperature. The execution, however, was extremely tedious, as will be evident.

The first step was to select a suitable material to form the emulsion. Apparently there had been some earlier studies; all that had been learned from them was that "a large number of colloidal solutions become clear in their upper portions when they are left alone for several weeks or several months." Perrin himself had made "some attempts without result on these colloidal solutions." Finally, "after some gropings," he had found it possible to make measurements on emulsions of gamboge, a yellow resin used in water colors,[7] or of mastic, a resin used in making varnish. Perrin does not specify the properties that he sought in his emulsions, but one feature that he is careful to mention about both gamboge and mastic is that the suspended particles are spherical. This meant that once a uniform suspension had been obtained, that is, one in which all the particles were the same size, the volume φ could be expressed in terms of the radius a, as $\varphi = \frac{4}{3}\pi a^3$—provided, of course, that the radius could be measured.

The suspensions as originally formed,[8] however, consisted of particles of a variety of sizes, and it was necessary to "sort them out" by size. This was done by a process that Perrin calls "fractional centrifuging." Here is his description of the technique:

> The centrifuge tube is filled to a given depth, 10 cm for example, with a pure emulsion; the machine is set in motion at a fixed angular speed, for example 30 revolutions per second (which gives, at 15 cm from the axis, a centrifugal force about 500 times greater than the weight); the drive of the machine is stopped after a fixed time, 60 minutes for example, and it is allowed to stop by itself, which must take a few minutes; the tube is carefully withdrawn.
>
> A fairly firm sediment, with a clearly defined surface, occupies the bottom of this tube to a depth usually negligible relative to the depth of the liquid; it contains all the grains that have reached the bottom during the centrifuging, pressed together much as the grains of sand filling a sack might be.
>
> Let us denote by a_1 the radius that a grain located at the surface of the liquid at the start of the centrifuging would have to have in order to arrive at the bottom of the tube just at the moment when the centrifuging stopped; every larger grain will *a fortiori* have arrived in the deposited sediment, but this sediment contains in addition many smaller grains, that had time to reach the bottom because they were found, at the beginning, at lower levels in the emulsion.
>
> By means of a siphon, the liquid which is above the sediment is carefully decanted; the tube is refilled with distilled water up to the original depth; it is shaken with the sediment, the grains of which all separate, and the preceding operation is carried out again with the

same angular speed and the same duration of centrifuging. All the grains of radius greater than a_1 will again have time to reach the bottom, but a small grain which previously was able to do so because it was already close will not be able to this time if chance makes it start near the surface. In short, the second sediment contains, like the first, all the grains whose radii exceed a_1, and contains many fewer smaller grains.

The supernatant emulsion is decanted, already paler than the fraction decanted first, and the same operations are repeated until the liquid that is found above the sediment at the end of each centrifuging becomes nearly clear water. Now this sediment contains all the grains of the original emulsion whose radius exceeds a_1 and no others; all the smaller grains have been eliminated.

Let us repeat the same operations on the final sediment, but with a little shorter duration of centrifuging. Let us designate by a_2 the radius that a grain at the surface must have to reach the bottom of the tube just at the end of this centrifuging. The liquid that is found above the sediment can only contain grains of a radius less than a_2; according to its origin, it can only contain grains of radius greater than a_1; thus *if* a_1 is near a_2, *this liquid is practically a uniform emulsion* which needs only to be decanted.

I think it is needless to explain how one can in an analogous fashion, if it is so desired, extract from the overlying fractions a uniform emulsion of still smaller grains, or from the residual sediment a uniform emulsion of still larger grains.

As is so often the case, the procedure appears simple and straightforward in retrospect. It was clearly ingenious, however, and must have been very time-consuming—note that Perrin gives no hint of how many repetitions were required for a complete separation.

The density of the grains was determined by two methods. One was simply to heat the emulsion so as to evaporate the water; the resin then melts, but on cooling solidifies again to a glassy substance whose density can be determined by standard methods. This procedure, however, implies the assumption that the resin has the same density in bulk form as it has in the form of suspended grains. Perrin therefore used a second method as a check. He determined the masses m of water and m' of emulsion needed to fill the same container at the same temperature. He then dried the emulsion and measured the mass μ of resin that it contained. If d is the density of water, the volume of the container is m/d;

the volume of the liquid originally in the emulsion is $(m' - \mu)/d$; the difference between these two numbers is then the volume occupied by the resin *in the emulsion,* which divided into its mass μ is its density. The two methods gave values in excellent agreement.

A flat cylindrical vessel was now formed by placing on a microscope slide a thin sheet of glass, usually 0.1 mm thick, with a hole through it.[9] A drop of uniform emulsion was placed in this vessel, and a microscope cover-glass sealed over it with paraffin or varnish. With evaporation thus prevented, a single preparation could be observed during a period of as much as several weeks. The whole was then placed on the stage of a microscope; it was necessary to take pains to see that the stage was level, so that motion of the body tube, and hence of the objective, would be accurately vertical. The whole arrangement is shown schematically in Fig. 4.1.

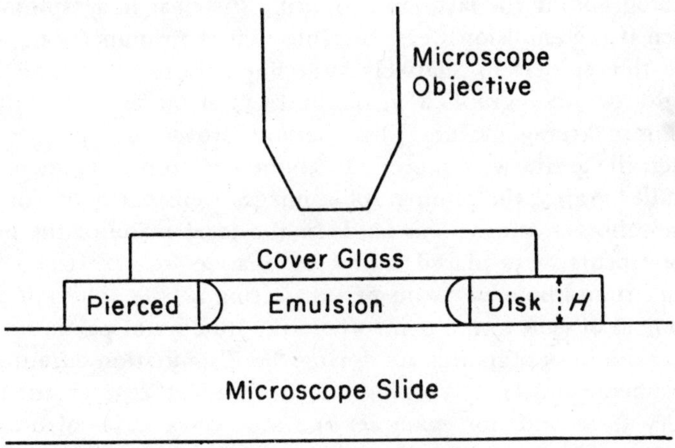

FIG. 4.1 Experimental arrangement for measuring the distribution of grains in an emulsion. The drawing proper is from Perrin's work; the legend has been translated. [Perrin, *Oeuvres Scientifiques*, p. 193, Fig. 21.]

The objective was chosen to give high magnification but to have only a small depth of field, of the order of a micron (the 0.1-mm depth of the container is equal to 100 microns). It was possible, then, to see clearly at one time only those grains in a very thin horizontal slab of the emulsion. If the body tube were raised or lowered, the grains in another slab would be seen, and the vertical distance between the slices, which would correspond to the difference in height h of Eq. (4.4), could be determined from the scale on the micrometer movement of the microscope.[10]

The next step was to determine the number of grains in a given area—the field of view, say—at each of two levels. This was by no means easy: ". . . it is not a matter of counting fixed objects, and when, on putting one's eye to the microscope, one discerns in the field some hundreds of grains that jiggle in every direction, and moreover continually disappearing at the same time as other grains appear, one is quickly convinced of the uselessness of any efforts that could be made to know even roughly the average number of grains present in the slice observed." This, of course, is an exaggeration. One means immediately suggests itself, that of photographing the layer and counting the clear images of grains. Even if the emulsion were so dilute that the number were small and thus subject to relatively large fluctuations,[11] the same layer could be photographed again and again so as to obtain an accurate average value. This method proved useful, but only when the grains were more than about $\frac{1}{2}$ micron in diameter; for smaller grains, the photographic images were not good. In such cases another scheme was used. In the focal plane of the microscope ocular was placed a disk of opaque foil, pierced with a small round hole by means of a dissecting needle. This cut down the field of view to the point where the number of particles could be taken in at a glance, say during the illumination admitted by a camera shutter. "Working in this way at regular intervals, every 15 seconds for example, one observes a series of numbers whose mean value approaches more and more closely a limit which defines the mean frequency of the grains, at the level studied, in the little cylindrical layer on which the microscope is focused." The same process repeated at a different level would give the mean frequency in an equal volume, and "the ratio of the two numbers gives the desired ratio of concentrations." It was necessary, however, to take several thousand readings to achieve any degree of precision.

The remaining quantity was the radius of the grains. Here again Perrin found more than one means of measurement, to provide a check. The first made use of Stokes's law, governing the fall of a sphere of radius a, of density Δ, through a fluid of density δ and viscosity η: The sphere acquires a limiting velocity called the terminal velocity, when the viscous drag of the fluid just counterbalances the net downward force, weight minus buoyancy; the terminal velocity satisfies the equation

$$6\pi\eta av = \tfrac{4}{3}\pi a^3(\Delta - \delta)g.$$

If an extremely tall vertical column of a uniform emulsion is produced, then, as Perrin puts it, "One will be far enough from the equilibrium distribution that the grains at the upper levels will sink like the droplets of a cloud, practically without one's needing to take into consideration the counter motion due to the accumulation of grains at lower levels. The upper part of the liquid thus becomes clear, and the extent of the clarified zone, divided by the time elapsed since the emulsion was left to itself, gives the velocity of fall to which Stokes's law is applied." The procedure, accordingly, was to seal a quantity of emulsion into a capillary tube (the narrow tube eliminated motion due to convection currents), install it vertically in a temperature bath, and observe the daily descent of the top of the cloud of grains; for such small grains, the terminal velocity would be reached in a matter of seconds, but would be measured in millimeters per day.

Perrin recognized that the application of Stokes's law to such small grains might be questionable,[12] so he used two more direct methods. One was to count the number of grains in a known volume of standardized emulsion—that is, emulsion in which the concentration of emulsion by mass was known. It was not practical, of course, to count the granules actually in suspension; but Perrin had "had occasion to notice that in a weakly acid medium (for example 0.01 gram mole per liter) the grains of gamboge or mastic collect on the glass walls that hold in the preparation." It was not a matter of destroying the suspension, but only that the grains stick to the glass when they happen to strike it, so that after several hours, all the grains originally present in one of his cells would be fixed to the surfaces of the container. Moreover, a "protective colloid" present in the natural latex of gamboge and added to the emulsion kept the grains separate rather than in partially coagulated globs. Finally, there was no significant lateral migration of the grains during the deposition process, so that those originally in a given vertical column ended up, apart from fluctuations, on the portions of the top and bottom plates that marked the ends of the column. Simple counting[13] then gave the number of grains in a known volume of emulsion, from which together with the mass per unit volume the size of each grain could be calculated.

A third method, still more direct, was usable for grains more than half a micron in diameter. In this case, it was found that the grains tended to arrange themselves in fairly regular arrays, within which could be distinguished straight rows of three to five grains. An example is shown in Fig. 4.2. While the image of a single grain was too much enlarged by diffraction to permit more than a rough estimate of its diameter, the length of such a row could be reasonably easily measured, either in direct view or on a photograph, thus permitting a determination of the diameter of a single grain.

The three methods gave closely comparable results. Six emulsions ultimately served as the source of Perrin's results; the values obtained for the radii of five of the six, and one extra, are given in Table 4-1. Perrin notes that to achieve the one case of three-

TABLE 4-1. *Radii of Grains in Six Emulsions, in Microns*

Material	Method Counting	 Stokes's law	 Rows
Gamboge	0.14	0.15	...
Gamboge	0.212	0.213	...
Gamboge	0.30	0.29	0.30
Gamboge	0.46	0.45	0.455
Gamboge	...	0.49	0.50
Mastic	...	0.52	0.54

figure accuracy, he had to count 11 000 grains.

The emulsion of 0.14-micron particles gave poor results, for reasons which Perrin does not specify. The resulting value of

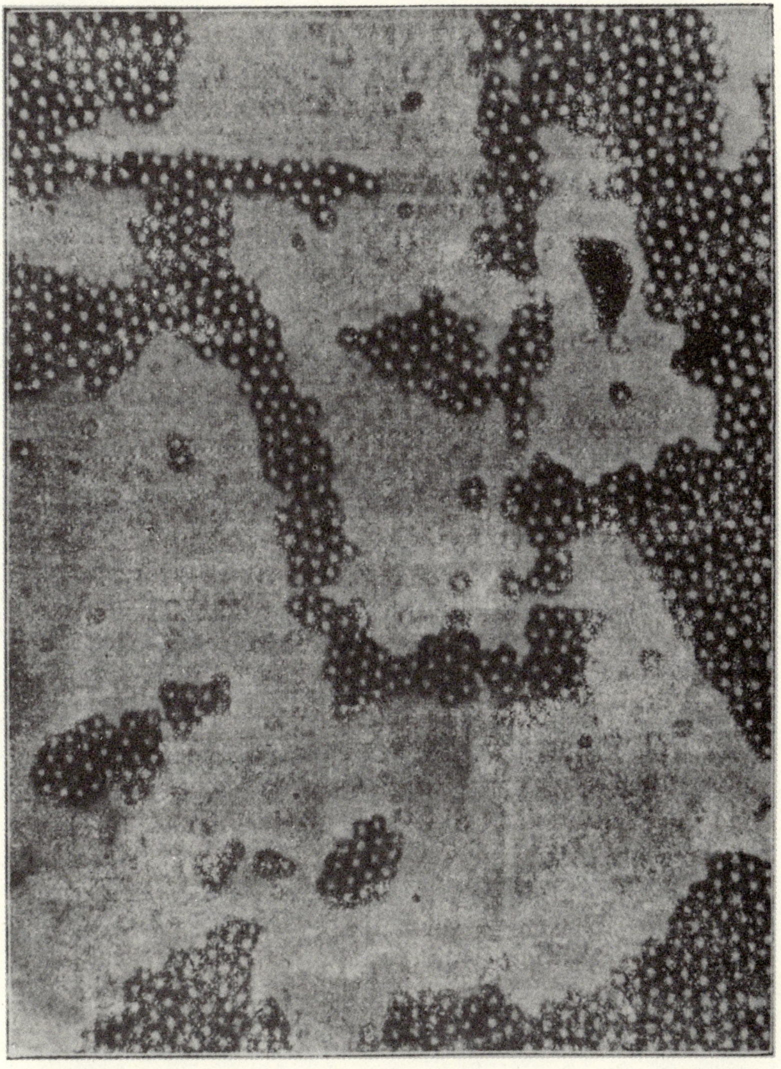

FIG. 4.2 A photograph of emulsion grains deposited on one surface of a container, showing the approximately regular array which permitted measurement of diameters. [Perrin, *Oeuvres Scientifiques*, facing p. 198, plate I.]

N_0 is given merely as "between" 5.0×10^{23} and 8.0×10^{23} per gram mole. Two series of measurements were made on emulsions of radius about 0.30 micron; one, probably the one whose radius measurements are tabulated, gave $N_0 = 7.5 \times 10^{23}$ per gram mole, while the other, for which the radius is reported as 0.29 micron, gave $N_0 = 6.6 \times 10^{23}$ per gram mole. The 0.45-micron emulsion gave $N_0 = 7.2 \times 10^{23}$ per gram mole, while the 0.52-micron one of mastic gave $N_0 = 7.0 \times 10^{23}$ per gram mole. The total numbers of grains counted in determining the concentration ratios for these five were between 3000 and 7500. The series of which Perrin was evidently proudest, however, was that on the 0.212-micron emulsion. He counted at four levels for a total of 13 000 grains, and obtained the result $N_0 = 7.05 \times 10^{23}$ per gram mole.

Considering the nature of the experiment and the difficulties involved, the agreement among these various values is quite satisfactory. Perrin notes that they cover a fortyfold range of masses, a nearly fivefold range of difference in density between grains and liquid,[14] and a thirtyfold range in rate of decrease of concentration with increase in height. He concludes, already at this point, that "it is quite difficult to deny the objective reality of molecules."

But he carried the matter still another step. It must not be thought that a given particle in Brownian motion merely undergoes an irregular motion around an equilibrium position. Rather, it traces out an irregular, wildly wandering path. Figure 4.3 shows three examples. On each of the paths, the points are plotted at intervals of 30 seconds; the size of the divisions is not given, but is of the order of one or a few microns. "One of these patterns contains 50 consecutive positions of a single grain. It gives only a very faint idea of the stupendous entanglement of the real trajectory. If, in fact, one were to make the mark second by second, each of the straight line segments would be replaced by a polygonal path of 30 sides, relatively as complex as the pattern here reproduced, and so on."

Einstein had analyzed this motion in a series of articles published in 1905 and 1906.[15] His analysis centered on the net displacement of a particle during an interval τ, that is, the distance from its position at the start of the interval to its position

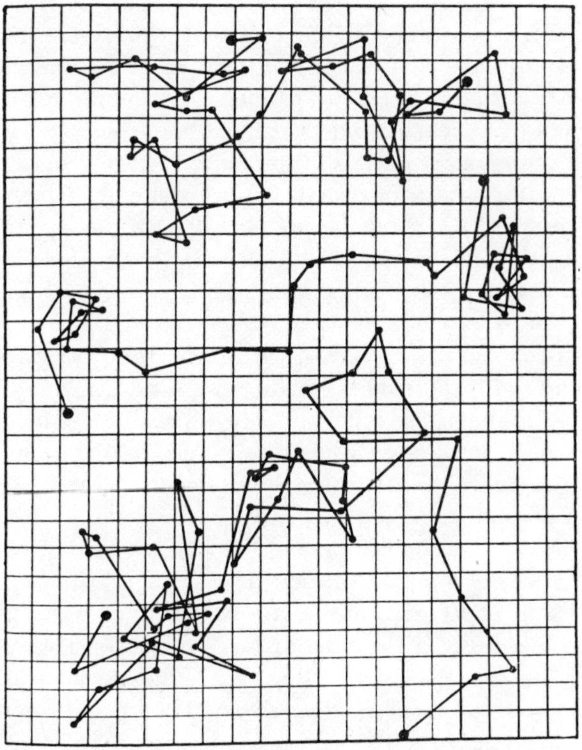

FIG. 4.3 Three examples of the paths of particles in Brownian motion. [Perrin, *Oeuvres Scientifiques*, p. 218, Fig. 25.]

at the end. As is evident from Fig. 4.3, this quantity does not have a single value, but many values. Each value is, of course, a vector quantity, for which the component along some specified direction can be found. What Einstein calculated was the average of the squares of the values of such a component. If this quantity is denoted by $\langle x^2 \rangle$, then for a spherical particle of radius a suspended in a liquid of viscosity η, the value is

$$\langle x^2 \rangle = \frac{RT}{N_0} \frac{\tau}{3\pi a \eta}, \tag{4.5}$$

where, as before, R is the gas constant, T is the absolute temperature, and N_0 is Avogadro's number.

In fact, at least three earlier studies had dealt with this aspect of Brownian motion; and all three, two of which were carried out after Einstein's work, appeared to be in contradiction with those results. Nevertheless, there were sufficient grounds for doubting the validity of the earlier works that Perrin and a student of his, named Chaudesaigues, were thoroughly willing to try again. Working with "grains of exactly known diameter which I knew how to prepare," Chaudesaigues "registered the position of a grain with the camera lucida,[16] from half-minute to half-minute, he began again with another grain, and so on. . . . There was then nothing more but to see whether the value given . . . by Einstein's equation . . . agrees, within the limits of error of the experiment, with the value already determined."

As a preliminary check, Chaudesaigues studied some grains of gamboge only indifferently identical. The results were encouraging, so a more careful effort was made. The 0.212-micron emulsion already discussed was used. Each of 100 grains was followed for four 30-second intervals, 50 grains at each of two values for the viscosity of the liquid. The first group of 50 gave $N_0 = 7.3 \times 10^{23}$ per gram mole, the second gave $N_0 = 6.8 \times 10^{23}$ per gram mole. For a third group of 50 grains the viscosity was increased about fivefold by adding sugar; the mean displacement was reduced in the theoretically predicted ratio, and the resulting value for N_0 was 5.6×10^{23} per gram mole.

Finally, three series of measurements were made, in collaboration with one M. Dabrowski, on emulsions of mastic. The values for N_0, in units of 10^{23} per gram mole, were 6.45, 7.15, and 7.7. The overall value obtained by this method was taken to be 7.15×10^{23} per gram mole, in quite reasonable agreement with the value determined from the distribution law.

Perrin devotes the last portion of his paper to a review of other methods of estimating N_0 (including a rather surprising one, due to Lord Rayleigh, involving the intensity of the blue light from the sky), and notes that all of them lead to results comparable with his own. He then gives this summary:

I believe it is impossible for an intellect free of prejudice to reflect on the extreme variety of phenomena that converge in this way toward the same result, without feeling a very strong effect; and I think that from now on it will be difficult to defend by rational arguments an attitude hostile to the molecular hypothesis, which breaks through all convictions one after another, and to which one must grant at least as much credence as to the principles of energetics.[17] There is certainly no question of contrasting these two great disciplines with one another, and the union of Atomistics and Energetics will glorify their double triumph.

FOOTNOTES

[1] All quotations in the present chapter are from this paper. The paper was of such significance that translations into English and German were published in 1910, the former being made by Frederick Soddy (see bibliography).

[2] All of the foregoing analysis may have been due to van't Hoff also; Perrin is, at times, careless about making clear how much of the background was drawn from the work of others. Similarly, the succeeding arguments may have been generally known.

[3] This is a far more drastic extension than it may appear at a casual glance. It ignores the fact that the original deduction depended on a distinction between solvent and solute.

[4] A single French word, *poussière,* is translated "grain of powder."

[5] The left-hand side is the difference between the two osmotic pressures, multiplied by the area on which each acts; the right-hand side is the difference of two terms each of which is a number of particles per unit volume times the volume $s\,dh$ of the slab, giving the total number of particles, times the volume of each particle, times a density, times the acceleration due to gravity, and is therefore the difference between the total weight of the particles and the total buoyant force on them.

[6] This step must be taken on faith by readers who have not studied calculus. The notation "ln" in Eq. (4.4) represents the logarithm to the base $e = 2.7182\ldots$, called "natural logarithm." Note that for any x, there holds $x = e^{\ln x} = (10^{\log e})^{\ln x} = 10^{\log e \ln x} = 10^{\log x}$, so that $\log x = \log e \ln x$. Conversely, $\ln x = \ln 10 \log x$.

[7] Perrin was chemist enough to be concerned over the chemical nature of the substances involved. For example, he established that the yellow substance which was suspended in the case of gamboge, and which constituted about 80% by weight of the raw material, was apparently a pure chemical compound rather than a mixture, that it was an acid (which he named *guttic* acid, from the French word for gamboge, *gomme-gutte*), and that it had a molecular weight in the neighborhood of 540.

[8] Commonly, by extracting the raw latex with alcohol and diluting the resulting solution with water.

[9] Such sheets are commercially available for use in making blood counts.

[10] A correction factor of $\frac{4}{3}$, allowing for the fact that what was being measured was an *apparent* depth of water, had to be applied if an immersion objective were not used; Perrin, however, usually used the immersion objective.

[11] If a single counting of a quantity subject to random fluctuation gives the result n, the probable error of this result, which is roughly equal to the mean fluctuation, is $\sqrt{n}$; the relative probable error is thus $1/\sqrt{n}$, which may be fairly large if n is only moderately large. For example, if the relative error is to be 1%, n must be 10 000.

[12] A few years later, Millikan was to discover that Stokes's law does indeed require modification when the diameter of the sphere becomes comparable with the average distance between molecules of the fluid. In liquids, however, this situation does not arise for visible, even microscopically visible, grains.

[13] In actuality, the positions of the grains were recorded on tracing paper by means of a camera lucida, an instrument which causes a virtual image to appear as if projected on a plane surface, such as a ground glass, so that a tracing can be made; in addition, the fluctuations mentioned were taken account of by averaging counts made on several vertical columns in a single cell, the columns being defined by square grills ruled on the microscope slide.

[14] For gamboge, the difference in density was in the range 0.2 to 0.3 gram per cubic centimeter, while for mastic it was only 0.063 gram per cubic centimeter.

[15] This work, and studies of the specific heat of solids, were at least as influential in the award of the Nobel prize to Einstein as was his work on relativity.

[16] The function of the camera lucida was described in footnote 13.

[17] The opponents of the atomic hypothesis, led by the chemist Ostwald, championed a doctrine that they called Energetics.

BIBLIOGRAPHY

This entire chapter is based on J. Perrin, *Annales de Chimie et de Physique* **18**, 1–114 (1909), reprinted in J. Perrin, *Œuvres Scientifiques* (Service des Publications du Centre National de la Recherche Scientifique, Paris, 1950), pp. 171–239. This important work was translated into both English and German. The English translation is by Frederick Soddy, *Brownian Movement and Molecular Reality* (Taylor and Francis, London, 1910), and is excerpted in *The World of the Atom,* pp. 628–640.

5 *The Atomic Nucleus*

In 1911, the atomic hypothesis was on firm ground as a consequence of Perrin's work. True, the experiments of Rutherford and Soddy had robbed the atom of its immutability; but radioactivity was a rather special phenomenon. Neither study, however, had revealed anything about the structure of the atom. Of that, little was known. It was generally accepted that "corpuscles" —what we now know as electrons—played some role in the structure; but what that role was remained unknown, except that it was certainly related to the spectrum of light emitted by the atom.

At least three different suggestions had been put forward, however. The earliest (1902) was due to Lord Kelvin, but was championed principally by Sir J. J. Thomson and has become known as the Thomson atom; Thomson published an extensive paper on it in 1904 and continued to work with it for several years. The model regards the positive charge and the mass of the atom as being distributed uniformly throughout the atom, presumably in a sphere whose diameter would be of the order of 1 angstrom (10^{-8} centimeter). Electrons were thought of as being embedded in this sphere like raisins in a pudding; in consequence, the model is frequently referred to as the "plum pudding" atom. Thomson studied possible equilibrium arrangements of the electrons in the atom, and noted that there could be expected a sort of periodicity of atomic properties that was at least qualitatively similar to what was observed.

The second was put forth by a professor at the University of Kiel, Philipp Lenard, who conceived it as a result of his work on cathode rays (work which earned him a Nobel prize) and published it in October 1903. He noted, first, that his work led to the conclusion that "the different atoms of all materials are constructed of constituents of one kind in different numbers." These constituents he called "dynamids." He concluded from the near-transparency of thin foils to cathode rays that the volume of the dynamids was only a tiny fraction—of order 10^{-12}—of the volume of the whole atom (which he, too, correctly estimated to be of the order of a few times 10^{-8} centimeter in linear dimension); and he suggested that the dynamid might consist of an intimate association of an electron and a positive body much more massive than the electron. However, he made no attempt to treat the dynamics of his model, and it was comparatively neglected (much to Lenard's embitterment).

The third model was the work of a Japanese, H. Nagaoka.[1] The full account was presented to the Physico-Mathematical Society of Japan and published in its *Proceedings*; but a sort of "extended abstract" was published in the British journal *Nature* in early 1904. The essential idea was that of a massive positive charge, with electrons spaced at regular intervals on a ring around it. The basic analysis of such a system had been carried out by Maxwell for the case of Saturn's rings; Nagaoka found that the repulsion between electrons, as contrasted with the attraction (gravitation) between the particles of Saturn's rings, did not alter the results. Nagaoka proposed that the small oscillations of the electrons about their equilibrium positions would give rise to optical radiation, and he achieved qualitative, and partially quantitative, agreement with the observed features of optical spectra. He also suggested that a large enough disturbance would disrupt the atom, with the electrons flying off as beta particles and the positive charge forming alpha particles. The model was promptly criticized on the grounds that it was highly unstable if it was electrically neutral. Nagaoka replied that his model was not intended to represent the entire structure of the atom, and so was not to be restricted to neutrality; he did not mention, however, what effects might be expected to result from the presence of the remaining electrons. His model, also, appears to have undergone little further development.

The decisive development came about as the result of a study of α-particle scattering, carried out by Hans Geiger, a postdoctoral fellow, and Ernest Marsden, an undergraduate student, under Rutherford's direction at the University of Manchester. They found that α particles are deflected through large angles with a probability too high to be explained except on the assumption that the atom contains a very small, massive core, the nucleus. This chapter describes their work.

The problem of α-particle scattering had attracted Rutherford's attention as early as 1906, while he was still at McGill University. He noted that while a beam of α particles collimated by means of apertures and traveling in air would produce a sharp-edged spot on a photographic plate, the spot would be blurred if a sheet of mica only 20 microns thick were interposed in the path. The blurring corresponded to a deflection of roughly 2 degrees for some of the particles. From considerations of the strength and extent of a magnetic field that would be needed to produce the same deflection, Rutherford deduced that the α particle must have experienced, while traversing the mica, an electric field of the order of 100 million volts per centimeter. "Such a result," he concluded, "brings out clearly the fact that the atoms of matter must be the seat of very intense electrical forces."

His attention was drawn to the problem again at Manchester in 1908, in the course of another study. He commended it to Geiger, who made a semiquantitative study reported in 1908. The apparatus used is shown in Fig. 5.1.

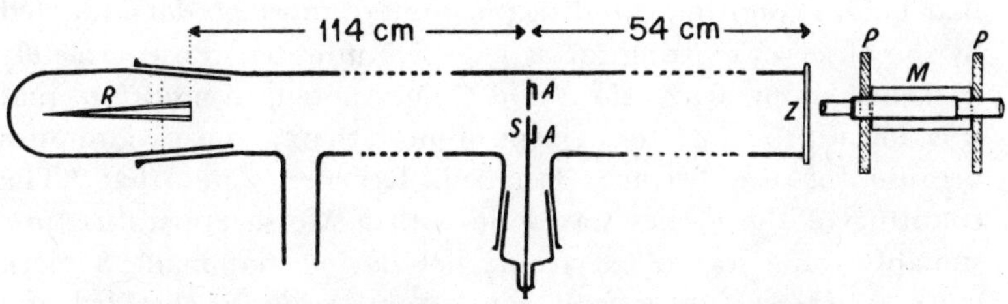

FIG. 5.1 Geiger's apparatus for studying α-particle scattering. [*Proc. Roy. Soc.* (London) *A81*, p. 174 (1908), Fig. 1; some lettering replaced for legibility.]

The main part consists of a glass tube nearly 2 metres in length and of about 4 cm. diameter. The α-particles from a strong but small source placed at R passed through a narrow slit S and produced an image of this slit on a phosphorescent screen Z, which was cemented to the end of the glass tube. The breadth of the slit was 0.9 mm., and the breadth of geometrical image on the screen was about 2 mm., depending upon the dimensions and distance of the source. The numbers of scintillations at different points of the screen were counted directly by means of a suitable microscope M, of 50 times magnification. The area of the screen which could be seen through the microscope was about 1 mm.[2] The number of scintillations counted varied between two or three a minute and about 80 per minute. . . . The microscope was mounted on a slide PP so that the scintillations produced at varying distances from the centre of the beam could be observed. The actual position of the microscope was read on a millimetre scale fixed to the slide.

The scintillation method, which here replaced the cruder photographic image of the earlier work and which was the detection system used in all the studies discussed in the remainder of this chapter, calls for some comment. The fact that zinc sulfide scintillates when struck by α particles had been known for some time; for example, Rutherford had mentioned it in a book on radioactivity published in 1904. Apparently the first person to adopt the phenomenon as a tool for research was an investigator in Berlin, Erich Regener, who reported his study in 1908. He noted that the method required careful choice of parameters of the microscope lenses to achieve maximum illumination from each α particle; that it is advisable to have the scintillating screen dimly illuminated by an ordinary lamp to provide a means for keeping the eye focused on the screen; and, most importantly, that before counting could begin, the eye must be dark-adapted by the observer's sitting for at least five minutes in a completely darkened room. Rutherford and Geiger noted, in addition, that it is difficult to count for periods of more than about two minutes because the eye becomes fatigued. Regener states that "The counting of the flashes was done with a Morse apparatus [presumably some sort of electromagnet device that made a mark when a key was depressed] and a stopwatch"; Rutherford and Geiger say, "It was usual to count 100 scintillations and to note the time with a stopwatch." Whatever the mechanics, the method was obviously both tedious and demanding of an extreme degree of concentration.

Geiger's results were presented in the form of the curves shown in Fig. 5.2.

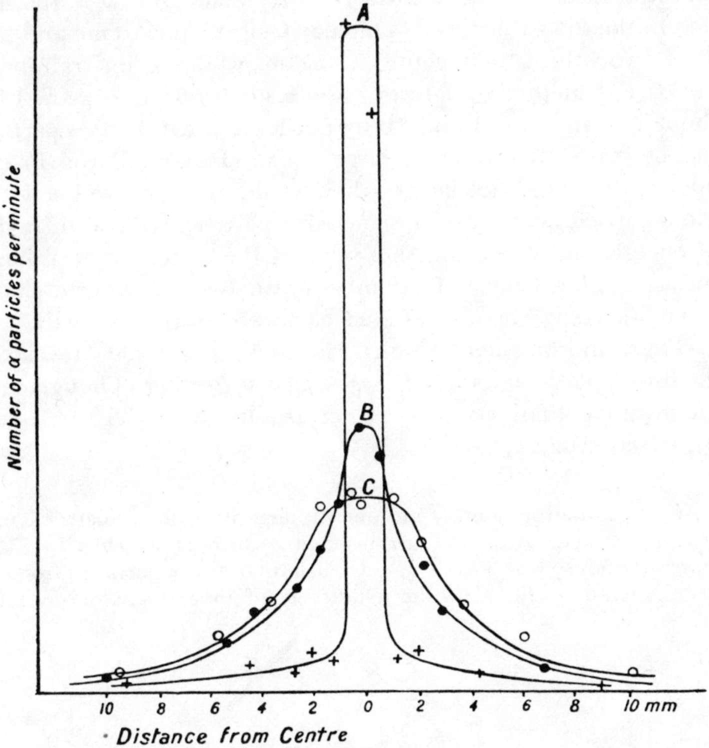

FIG. 5.2 Geiger's curves of α-particle scattering. [*Proc. Roy. Soc. (London) A81*, p. 176 (1908), Fig. 2.]

The curve *A* shows the distribution of the scintillations in a charcoal vacuum.[2] . . . The second curve *B* shows the effect if the slit is covered with one gold leaf.[3] The area over which the scintillations were observed was much broader and the difference in the distribution could easily be noticed with the naked eye. . . . The third curve *C* shows the effect of two gold leaves together.

It will be noted that Geiger's data extended only to deflections of some 10 millimeters in a distance of 54 centimeters, corresponding to angles of only a degree or so. Even so, the number of counts had dropped markedly; very few particles were scattered through angles larger than that.

Rutherford, in a lecture nearly thirty years later, told the next step in the story this way:[4] "One day Geiger came to me and said, 'Don't you think that young Marsden, whom I am training in radioactive methods, ought to begin a small research?' Now I had thought that too, so I said, 'Why not let him see if any α-particles can be scattered through a large angle?' I may tell you in confidence that I did not believe they would be, since we knew that the α-particle was a very fast massive particle, with a great deal of energy, and you could show that if the scattering was due to the accumulated effect of a number of small scatterings the chance of an α-particle's being scattered backwards was very small."

Geiger and Marsden's results, reported in June 1909, revealed a startlingly large amount of large-angle scattering. Though more quantitative than Geiger's earlier results, they were still comparatively crude.

> . . . as radiating source, radium C, deposited on a plate of small dimensions, was used. The arrangement, which is sketched in Fig. 3 [here reproduced as Fig. 5.3], was such that the α-particles from the plate A fell upon the platinum reflector R, of about 1 square centimetre

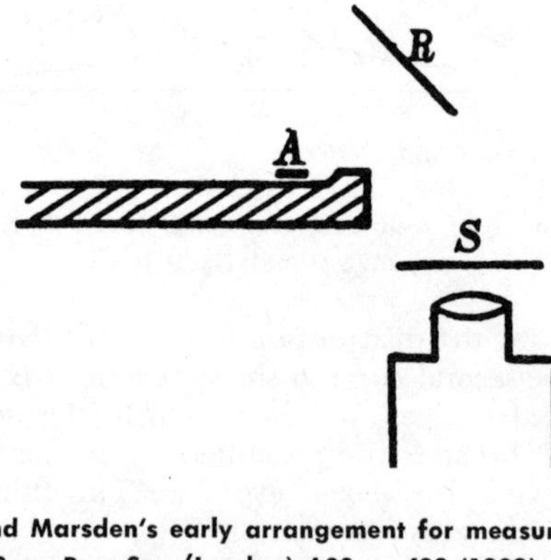

FIG. 5.3 Geiger and Marsden's early arrangement for measuring α-particle scattering. [*Proc. Roy. Soc.* (London) *A82*, p. 499 (1909), Fig. 3.]

area, at an average angle of 90°. The reflected particles were counted on different points of the screen S. . . .

The amount of radium C deposited on the plate was determined by its γ-ray activity. Assuming that 3.4×10^{10} particles are expelled per second from an amount of RaC equivalent to 1 gramme Ra,[5] the number of α-particles expelled per second from the active plate was determined. The number falling on the platinum reflector was then easily calculated from its known distance and area. To find the whole number of reflected particles, it was assumed that they were distributed uniformly round a half sphere with the middle of the reflector as centre.

Three different determinations showed that of the incident α-particles about 1 in 8000 was reflected, under the described conditions.

In the lecture already referred to,[6] Rutherford described his reaction to learning of this result:

> . . . I remember . . . Geiger coming to me in great excitement and saying, 'We have been able to get some of the α-particles coming backwards. . . .' It was quite the most incredible event that has ever happened to me in my life.[7] It was almost as incredible as if you fired a 15-inch shell at a piece of tissue paper and it came back and hit you. On consideration I realized that this scattering backwards must be the result of a single collision, and when I made calculations I saw that it was impossible to get anything of that order of magnitude unless you took a system in which the greater part of the mass of the atom was concentrated in a minute nucleus. It was then that I had the idea of an atom with a minute massive centre carrying a charge.[8]

Rutherford evidently recognized that according to the Thomson model of the atom, the incident particle would not experience a very large force as the result of a collision with an atom, and therefore could not undergo a very large deflection in a single collision. Rather, a large deviation would have to be the cumulative effect of a number of successive small deflections. But although an α particle would have a large chance of being multiply scattered in passing through matter composed of Thomson atoms, it is very unlikely that the many deflections would be all in the same direction. A large net deflection, therefore, was extremely unlikely.

Apparently, though he was not much of a mathematician nor theorist, Rutherford also recognized that if the positive charge were concentrated in a very small region, the large Coulomb force required to make such a large deflection possible in a single collision could be experienced; at the same time, while the likelihood of collision taking place was much reduced, the likelihood of multiple scattering (at least in a thin foil) was much more so. He worked out the details (published in 1911),[9] and found that the likelihood of an α particle being deflected through an angle θ in passing through a foil of thickness t is proportional to $t \csc^4(\frac{1}{2}\theta)$. For the Thomson model, on the other hand, the variation with foil thickness is as $t^{1/2}$ and the angle dependence is dominated by a factor $\exp(-\theta^2/\theta_{av}^2)$, where θ_{av} is the average angle of deflection—of the order of a degree or so. The differences were susceptible to experimental test, but the experiment had to be much more precise than the preceding work had been. Rutherford's paper concluded with the statement, "Experiments in this direction are already in progress by Geiger and Marsden."

The results were published in 1913. The apparatus was designed to meet the condition that "to obtain measurable effects an intense pencil of α-particles is required." As shown in Fig. 5.4, it

> . . . mainly consisted of a strong cylindrical metal box *B*, which contained the source of α particles *R*, the scattering foil *F*, and a microscope *M* to which the zinc-sulphide screen *S* was rigidly attached. The box was fastened down to a graduated circular platform *A*, which could be rotated by means of a conical airtight joint *C*. By rotating the platform the box and microscope moved with it, whilst the scattering foil and radiating source remained in position, being attached to the tube *T*, which was fastened to the standard *L*. The box *B* was closed by the ground-glass plate *P*, and could be exhausted through the tube *T*.
>
> The source of α particles . . . consisted of a small thin-walled glass tube about 1 mm. in diameter, containing a large quantity of well purified radium emanation. The α particles emitted by the emanation and its active deposit could pass through the glass without much reduction in range. For these experiments the inhomogeneity of the source, due to the different α particles from the emanation, Ra A and Ra C, does not interfere with the application of the law of scattering with angle as deduced from the theory, as each group of α particles is scattered according to the same law.
>
> By means of a diaphragm placed at *D*, a pencil of α particles was directed normally on to the scattering foil *F*. By rotating the microscope the α particles scattered in different directions could be observed on the

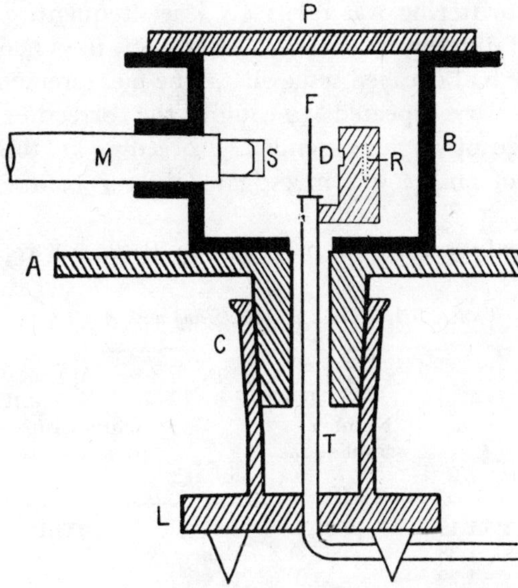

FIG. 5.4 Geiger and Marsden's later apparatus for measuring the angular variation of α-particle scattering. [*Phil. Mag. 25*, p. 607 (1913), Fig. 1.]

screen *S*. . . . Observations were taken in various experiments for angles of deflexion from 5° to 150°. When measuring the scattering through large angles the zinc-sulphide screen had to be turned very near to the source, and the β and γ rays produced a considerable luminescence on it, thus making countings of the scintillations difficult. The effect of the β rays was reduced as far as possible by a lead box shown shaded in the diagram. The amount of lead was, however, limited by considerations of the space taken up by it, and consequently observations could not be made for angles of deflexion between 150° and 180°. . . .

Even when no scattering foil was used a few scintillations were always observed on the screen. They were obviously due to scattered radiation from the walls of the vessel and from the edge of the diaphragm limiting the beam. The effect was reduced as far as possible. . . . The number of stray α particles was determined for different positions of the microscope by removing the scattering foil so that the necessary corrections could be applied with certainty.

The customary procedure was as follows: While the source was fresh and strong, a set of measurements was made at large angles, at which the scattering was relatively least frequent, on each of several foils of different materials. "After a few days had elapsed" and the source had decayed somewhat, "the measurements for the smaller angles were repeated" to confirm the correction for decay "and the range of angles extended. Proceeding in this way the whole range of angles was investigated in the course of a few weeks."

The principal results are reproduced in Table 5-1, representing

TABLE 5-1. *Variation of Scattering with Angle*

I	II	III	IV	V	VI
Angle of deflexion, φ (deg)		SILVER		GOLD	
	$\frac{1}{\sin^4 \varphi/2}$	Number of scintillations, *N*	$\frac{N}{\sin^4 \varphi/2}$	Number of scintillations, *N*	$\frac{N}{\sin^4 \varphi/2}$
150	1.15	22.2	19.3	33.1	28.8
135	1.38	27.4	19.8	43.0	31.2
120	1.79	33.0	18.4	51.9	29.0
105	2.53	47.3	18.7	69.5	27.5
75	7.25	136	18.8	211	29.1
60	16.0	320	20.0	477	29.8
45	46.6	989	21.2	1435	30.8
37.5	93.7	1760	18.8	3300	35.3
30	223	5260	23.6	7800	35.0
22.5	690	20300	29.4	27300	39.6
15	3445	105400	30.6	132000	38.4
30	223	5.3	0.024	3.1	0.014
22.5	690	16.6	0.024	8.4	0.012
15	3445	93.0	0.027	48.2	0.014
10	17330	508	0.029	200	0.0115
7.5	54650	1710	0.031	607	0.011
5	276300	...	...	3320	0.012

two series of measurements on silver foils of equal thickness, and two on gold foils of unequal thicknesses.

. . . Col. I gives the values of the angles φ between the direction of the beam and the direction in which the scattered α particles were counted. Col. II gives the values of $1/\sin^4\varphi/2$. In Cols. III and V the numbers of scintillations are entered which were observed for the silver and gold respectively. Corrections are made for the decay of the emanation, for the natural effect, and for change of diaphragm. . . .

In Cols. IV and VI the ratios of the numbers of scintillations to $1/\sin^4\varphi/2$ [*sic*] are entered. It will be seen that in both sets the values are approximately constant. The deviations are somewhat systematic, the ratio increasing with decreasing angle. However, any slight asymmetry in the apparatus and other causes would affect the results in a systematic way so that . . . the deviations from constancy of the ratio are probably well within the experimental error. The experiments, therefore, prove that the number of α particles scattered in a definite direction varies as $\operatorname{cosec}^4\varphi/2$.

The test of the dependence on thickness required a somewhat different apparatus, shown in Fig. 5.5.

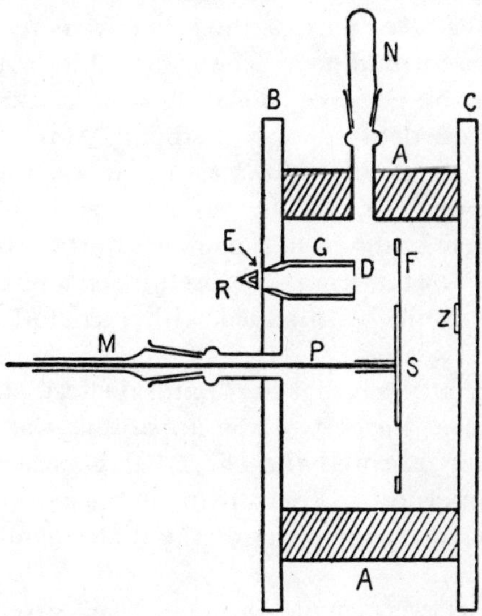

FIG. 5.5 Geiger and Marsden's apparatus for measuring the thickness dependence of α-particle scattering. [*Phil. Mag.* 25, p. 612 (1913), Fig. 3.]

It consists essentially of a source of a radiation *R,* a diaphragm *D,* a scattering foil *F,* and a zinc-sulphide screen *Z* on which the scattered α particles were observed. The main part of the apparatus was enclosed in a cylindrical brass ring *A,* the ends of which were planed so that they could be closed airtight by the two glass plates *B* and *C.* The depth of the ring was 3.5 cm., and its internal and external diameters 5.5 and 7.5 cm. respectively. Two holes were drilled through the glass plate *B,* one in the centre and the other 1.65 cm. excentric. The source of radiation *R* was placed directly against a sheet of mica which was waxed over and closed the opening *E*. . . .

By means of the diaphragm *D* a narrow pencil of α particles could be directed on to the scattering foil. The different foils were attached to the disk *S* and covered five of six holes drilled through it at equal distances from the centre. The uncovered opening was used to determine the natural effect. The disk could be fitted on to the rod *P*, which was fastened to the ground-glass joint *M* so that it could be rotated and the different foils brought in front of the diaphragm. The scattered α particles were observed by means of a microscope on the zinc-sulphide screen *Z* fixed inside the glass plate.

The preparation of the source required rather more elaborate care than for the other part, as the inhomogeneity that was of no significance there would have been intolerable here. It was found possible to prepare a source consisting of a mixture of Ra A and Ra C. The former decayed away in about 20 minutes after preparation, leaving essentially pure Ra C; this was the actual source and was usable for about another hour, after which small amounts of emanation remaining from the mode of preparation interfered. The resulting short intervals of usefulness limited the number of counts that could be obtained with each foil and hence the accuracy.

Another difficulty was that the "natural effect," the background of scattering from the rest of the apparatus, was so strong that in some cases it obscured the effect being measured. This was overcome by interposing a screen in such a position as to block particles scattered by the edges of the diaphragm from reaching the screen.

Measurements were made on gold, tin, silver, copper, and aluminum. The results were plotted as numbers of scintillations per minute, corrected for variation of strength of the source with time, versus foil thickness. "For all the metals examined the points lie on straight lines which pass through the origin. The experiments therefore prove that for small thicknesses of matter the scattering is proportional to the thickness."

Rutherford's theory also predicted that the scattering of α particles of different velocities would be inversely proportional to the fourth power of the velocity. Geiger and Marsden tested this part of the theory also, and found it to be correct. Thus, they concluded, "we have completely verified the theory given by Prof. Rutherford." The nuclear atom was established as a reality.

FOOTNOTES

[1] The average Occidental may be startled to find that the Japanese were taking a significant part in the scientific developments of the early 1900's. The fact is that beginning in 1868, the government of Japan had strongly encouraged the growth of science and the participation of Japanese in research.

[2] That is, a vacuum in which the pressure is reduced by adsorption of gas on charcoal (often in a cold trap), to the order of 10^{-4} millimeter of mercury (10^{-7} atmosphere).

[3] Thickness not given. Standard gold leaf is between 4 and 5 millionths of an inch thick.

[4] *Background to Modern Science,* edited by Joseph Needham and Walter Pagel (Cambridge University Press, London, 1938), pp. 68–69.

[5] This number had been determined earlier by Rutherford and Geiger. It had been that study that stimulated the further work on α-particle scattering.

[6] See footnote 4.

[7] This remark whets the reader's curiosity as to Rutherford's reaction to the discovery of transmutation (Chap. 3).

[8] How much Rutherford was influenced by Nagaoka's work, to which his bears some obvious resemblances, is impossible to determine. He certainly knew of it—his paper on the details of the theory contains a reference to it; and it is quite possible that he had discussed it with Nagaoka when the latter visited Manchester in 1910.

[9] Curiously, the word "nucleus" does not appear in this paper. Nor is there any explicit mention of the assumption that the mass of the atom is concentrated along with the charge, though this is implicit in the dynamical considerations.

BIBLIOGRAPHY

The Thomson atom is treated in J. J. Thomson, *Philosophical Magazine* **7,** 237–265 (1904); the Nagaoka model is sketched in H. Nagaoka, *Nature* **69,** 392–393 (1904). The latter reference is included because of its possible influence on Rutherford's work.

The preliminary experimental work is reported in H. Geiger, *Proceedings of the Royal Society of London, Series A* **81,** 174–177 (1908); H. Geiger and E. Marsden, *ibid.* **82,** 495–499 (1909). The latter is excerpted in *The World of the Atom,* pp. 696–700.

The theoretical work was published by E. Rutherford, *Philo-*

sophical Magazine **21**, 669–688 (1911). It is reprinted in Sir James Chadwick, ed., *The Collected Papers of Lord Rutherford of Nelson* (George Allen and Unwin Ltd., London, 1963), Vol. 2, pp. 238–254, and is excerpted in *The World of the Atom*, pp. 707–722.

The definitive experimental work is presented in H. Geiger and E. Marsden, *Philosophical Magazine* **25**, 604–623 (1913), and is excerpted in *The World of the Atom*, pp. 722–733.

6 *Collisions of Electrons with Atoms*

The properties of atoms continued to be investigated with interest. Among other things, it was known that if sufficient energy is supplied to an atom, one (or more) electron may be knocked loose from the atomic structure, changing the atom into an ion. When the energy is supplied by striking the atom with an externally produced electron, it is measured in terms of the potential difference through which the electron has been accelerated. The value of this potential difference necessary to produce ionization is called the ionization potential, and is of significance in its own right in connection with the understanding of phenomena of electrical discharge in gases. Consequently, a number of determinations of ionization potentials had been made, for various gases, in the early years of this century. They were mostly indirect, however, dependent on uncertain hypotheses, and the values obtained for a given substance by different workers might differ widely.

Among the investigators in this area were James Franck and Gustav Hertz. Working at the University of Berlin, they measured what they thought to be the ionization potentials of several elements. It later became clear that they had not actually measured an ionization potential; rather, their work, for which they won the Nobel prize in 1925, revealed that there was a minimum amount of energy that could be absorbed by a given variety of atom, and that this minimum energy was related to the optical spectrum of the element.

Their initial method was to produce electrons by means of a heated filament; accelerate them through a measured, variable potential difference V maintained between the filament and a grid; and decelerate them again by another, reverse potential difference, which was ten volts greater than V, between the grid and a collector plate. In the space between grid and collector, the electrons could collide with gas atoms. They could not under any circumstances reach the collector, because of the second potential difference. However (the argument ran), if V were greater than the ionization potential, the electrons would ionize some atoms by collision; the electric field would accelerate the positive ions toward the collector, which would then register a current. The value of the ionization potential, then, was the value of V for which current began to flow to the collector. By this method, Franck and Hertz measured what appeared to be ionization potentials for a half-dozen gases.

Their intention was to test a proposed correlation of the ionization potentials with atomic radii. For this purpose they wanted to make measurements on metallic atoms, which have relatively large radii. They felt that maintaining their apparatus at the higher temperatures needed to produce metal vapors would cause errors—presumably, the currents they worked with were so small that the decrease in the resistance of the glass envelope, produced by the increase in temperature, would give rise to stray currents large enough to mask the desired effect. Accordingly, they devised a new form of apparatus, shown in Fig. 6.1. The grid N, instead of being fairly close (5 mm) to the filament, was made about 4 cm away; the collector plate G was placed only 1 or 2 mm from the grid, instead of some 2.5 cm. The potential between filament and grid was variable, and such as to accelerate the electrons, as before; the potential between grid and collector was again decelerating, but now small and constant.

The operation of the apparatus was described by Franck and Hertz in the report on their work, published in 1914: "As long as the accelerating potential is less than the decelerating, the current [to the collector] is null. Then it will increase, until the accelerating potential has become equal to the ionization potential." The basis for this statement is as follows: The electrons could

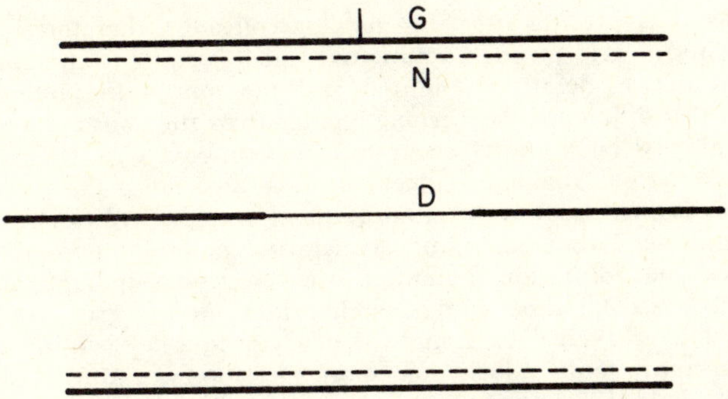

FIG. 6.1 Electrode arrangement for Franck and Hertz's measurements of presumed ionization potentials. [*Verhandl. Deut. physik. Ges.* 16, p. 459 (1914), Fig. 1.]

collide with gas atoms in the accelerating region. But it was assumed that a collision between an electron and an atom would be elastic[1] as long as the kinetic energy of the electron is less than the ionization energy; and in an elastic collision with an atom, the electron loses a negligible fraction of its energy[2] and thus can still overcome the opposing effect of the second potential and reach the collector. The assumption involved will be discussed later, but it seemed reasonable in the light of the earlier experiments. The account goes on:

> At this point, the electrons will undergo inelastic collisions in the neighborhood of the grid and thereby ionize. Since they themselves and the electrons set free by ionization traverse only a very small potential until their passage through the grid, they pass through it without an appreciable velocity and are incapable of running against the retarding field. The galvanometer current will, therefore, drop to zero as soon as the accelerating potential has become greater than the ionization potential.

This implies a second assumption: that the probability of a collision being inelastic, if the electron energy is large enough for that to be possible, is not very small compared to unity. This assumption, also, seemed to be borne out by the earlier work. Franck and Hertz continue:

> If one increases the accelerating potential further, the point at which the electrons undergo inelastic collisions moves inward from the grid.

The electrons present after the inelastic collisions, therefore, pass on the way to the grid through a potential which is equal to the difference between the accelerating potential and the ionization potential. As soon as this difference has become greater than the constant retarding potential between N and G, electrons can again go against the retarding field and the galvanometer current again grows. Since the number of electrons is increased by the ionization, it actually grows larger than the first time. However, as the accelerating potential becomes equal to twice the ionization potential, the electrons undergo inelastic collisions a second time in the neighborhood of the grid. Since they thereby lose all their energy, and the newly produced electrons likewise have no appreciable velocity, no more electrons can run against the retarding field. Thus, as soon as the accelerating potential is greater than twice the ionization potential, the galvanometer current again drops to zero. Since the same phenomenon repeats itself each time the accelerating potential is equal to an integral multiple of the ionization potential, we should expect a curve having maxima of increasing size, whose separation is equal to the ionization potential.

Just such curves were obtained; an example is reproduced in Fig. 6.2. The maxima turned out to be quite sharp—the report gives a feeling that they were sharper than the authors had dared to expect—and Franck and Hertz imply a confidence that their results are accurate to 0.1 volt as compared with the 1-volt accuracy of the older method. The value obtained for the ionization potential of mercury was 4.9 volts. As a comparison with the older method, they remeasured the value for helium and found that the two methods gave very satisfactory agreement.

Such accuracy exceeded anything previously obtained, and enabled Franck and Hertz to make a quantitative test of a theoretical proposal that had been put forth several times: that the ionization energy should be equal to Planck's constant h times the frequency of one of the "proper motions" of the electrons. They felt it natural to choose a frequency that was very strongly absorbed by mercury vapor, that corresponding to a wavelength of 2536 Å. The potential indicated by the theory was 4.84 volts, in excellent agreement with the measured value.

At this stage, perhaps out of the very excellence of the agreement, doubts began to arise. The possibility presented itself that the electrons lost their energy not in ionizing the mercury—after all, in this method ionization had not been directly observed—but in exciting radiation. They could not check this new possi-

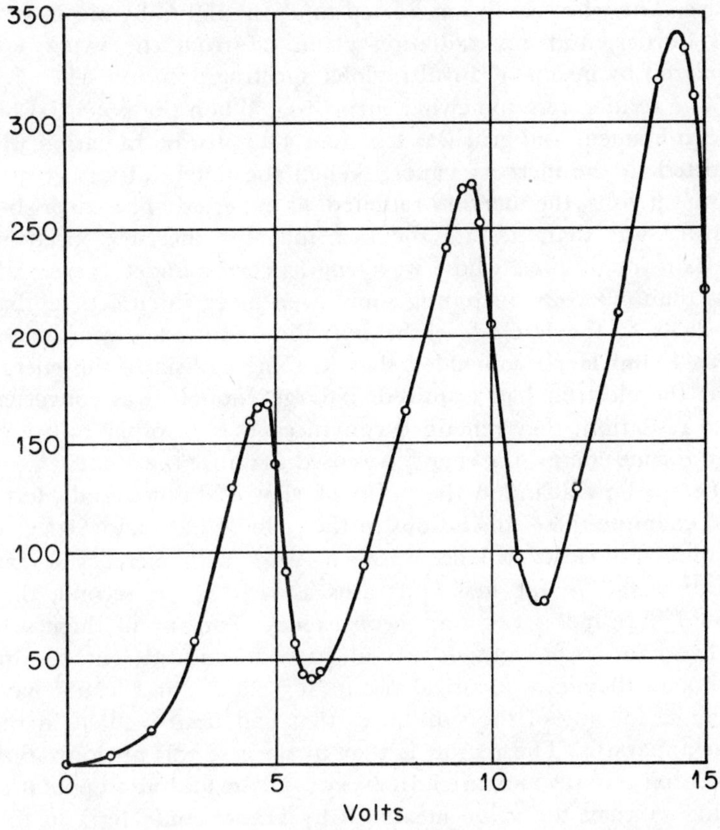

FIG. 6.2 Plot of collector current versus accelerating potential, showing the equal spacing of the maxima. (*Note:* Several effects combine to cause the spacing between maxima to be slightly different from the position of the first maximum.) [*Verhandl. Deut. physik. Ges. 16*, p. 462 (1914), Fig. 3; retouched for adequate reproducibility.]

bility with the same apparatus, as the wavelength 2536 Å lies well into the ultraviolet, while the glass envelope of their tube was opaque to ultraviolet light. Consequently, they built a new tube out of quartz, which is transparent to ultraviolet light. This tube was much simpler than the others, involving merely a platinum filament to provide electrons, a platinum grid toward which the electrons could be accelerated and on which they were collected, and a pool of mercury to provide vapor in the intervening space. The whole bulb was heated to about 150°C by means of a gas burner, and any radiation given off from the vapor was analyzed by means of an ultraviolet spectrograph.

The results were somewhat surprising. When the potential between filament and grid was less than 4.9 volts, no radiation was emitted by the mercury vapor.[3] When the potential was greater than 4.9 volts, the mercury radiated, as expected. The surprising feature was that, despite the fact that the mercury spectrum contains many lines whose wavelengths correspond to potentials less than 4.9 volts, including some even more intense than that at 2536 Å, the latter was the *only* line emitted. Nevertheless, Franck and Hertz concluded that in some collisions, the energy that the electron had acquired, if large enough, was converted into radiation; they remained convinced that in other collisions at the same energy, the energy was used to ionize the atom.

Let us now abandon the point of view of Franck and Hertz, and examine these conclusions in the light of later knowledge. If Franck and Hertz had been able to work with mercury (or an alkali metal) in the first apparatus as well as the second, they would have met a peculiar inconsistency: For any of these substances, the two methods would have given different results, although they gave identical results for helium and would have done so for any of the substances that had been studied in the first apparatus. The reason is that in no case was an ionization potential actually measured. In every case, the ionization potential is higher than the value measured by Franck and Hertz in this work. The increases in collected current which Franck and Hertz ascribed to positive ions had a much different origin. The atoms or molecules were being raised to a higher energy state by the incident electrons and were emitting radiation, and the radiation was producing photoelectrons from the collector plate. The radiation emitted by mercury, and the radiation that would have been emitted by an alkali metal, could not have produced this effect.[4]

What, then, do the measured values mean? It is true that when the current to the collector drops, it is a signal that the electrons have gained just enough energy to lose most of it in what must be an inelastic collision with an atom, one that alters the *internal* behavior of the atom. When it drops a second time, the electrons have gained enough energy to experience such an encounter twice during their travel from filament to grid, and so on. The data show that there is a minimum amount by which the internal energy of the atom can be changed—and in that statement, forced on us by the experimental facts, is found the failure of extrapolation from experience on the ordinary scale. On the ordinary scale of experience, it is perfectly acceptable for a dynamical system to have characteristic *frequencies,* as an atom does; a pendulum or a stretched string comes immediately to mind, and other more complicated examples could be cited. But the *energy* of any such system can be changed by an arbitrary amount. Apparently, the same is *not* true of an atomic system. It can exist only in certain states with certain, discrete energy values. Its state can change only from one of these to another, and so its energy can change only by certain distinct amounts. This is the lesson of the Franck-Hertz experiment. And even as the experiments were being done, Niels Bohr was taking this idea as one of the basic postulates on which to build his theory of the atom, quite unaware that his radical proposal was being verified.

FOOTNOTES

[1] The reader is reminded that collisions are classed as elastic or inelastic according to what becomes of the initial kinetic energy. If it remains as kinetic energy of the two colliding bodies, though perhaps shared differently, the collision is elastic. If some of it is absorbed so as to alter the internal state of one of the colliding bodies, then the collision is inelastic.

[2] The reader can verify that if a body of mass M, initially at rest, is struck by a body of mass m, and if the collision is elastic, the incident body loses a fraction of its energy which is at most $4mM/(M+m)^2$; when $m \ll M$, this fraction is very nearly $4m/M$.

[3] The critical value in this arrangement actually turns out to be somewhat less than 4.9 volts because the electrons already have some energy when they are produced by the filament.

[4] For a further discussion realting to this point, see Chap. 7.

BIBLIOGRAPHY

J. Franck and G. Hertz, *Verhandlungen der Deutschen Physikalischen Gesellschaft* **16**, 457–467 (1914) [translation: *The World of the Atom,* Vol. 1, pp. 770–778].

7 *The Photoelectric Effect*

We have seen in Chapter 2 that the quantum concept was originally introduced as an aspect of the behavior of radiating oscillators rather than of the radiation itself. It was natural, however, to feel that such behavior might impress itself at least partially on the radiation. The quantum concept had existed only five years when Albert Einstein seized upon this possibility as an explanation for the peculiar effects of radiation in the photoelectric effect. It was not until 1914, however, that a thorough, convincing test of Einstein's ideas was reported.

The photoelectric effect was discovered late in the nineteenth century, and by 1914 had been studied fairly extensively. The basic phenomena were known: A beam of light striking the surface of a metal liberates electrons from the metal, provided its frequency is greater than a critical value dependent on the kind of metal. The liberated electrons emerge with some kinetic energy, the amount of which increases with increasing frequency of the light according to a relationship whose form had not been experimentally established with certainty in 1914. If the electrons are collected at another electrode and made to constitute a current, the magnitude of the current is proportional to the intensity of the stimulating light. The entire process is virtually instantaneous.

Almost every aspect of the photoelectric effect appears as a failure of extrapolation. The phenomenon was discovered and some of its properties established, including at least crude recognition of critical frequencies, before the electron was discovered; but there was nothing in classical theory, dealing as it did with continua, that would have suggested such a process. Even given its existence as an empirical fact, the only property that could be reasonably contained within the classical framework was the dependence of the photocurrent on the light intensity. Once the electron was discovered, it was possible to envision an interaction between electrons in the metal and the electric field of the light wave, but this only made matters worse. The only mechanism that could be reasonably envisioned by which an electron would be extracted from the metal was through a resonance in which the frequency of the light matched a natural frequency of vibration of the electron. This, however, would lead to photoemission only for discrete frequencies, or at most fairly well-defined, isolated bands of frequencies, rather than throughout a continuous range; and it would not explain the dependence of the electron energy on frequency. The time dependence is an even more drastically startling result. The assumption of a resonance process for the transfer of energy from the light to the electron implies that the electron will absorb all the energy incident on an area one wavelength square. Computations on this basis lead to the conclusion that for a beam of very low intensity, but still sufficient for the photoelectric effect to be easily observed, an electron would take about 500 *years* to accumulate energy equal to that observed.

Einstein's proposed explanation of the phenomenon was based on a radical extension of Planck's original quantum hypothesis. That hypothesis, it will be recalled, was that the energy of an oscillator was quantized; by implication, the energy radiated into the electromagnetic field by the oscillator would also be quantized. Einstein suggested that such a "bundle" of energy maintained its unity in the process of absorption also, specifically in its absorption by an electron in a metal. The consequence was a simple equation relating the kinetic energy E_k of the photoelectrons and the frequency ν of the stimulating radiation:

$$E_k = h\nu - e\varphi, \tag{7.1}$$

where h is Planck's constant and φ is a constant characteristic of the metal from which the electrons were ejected.

Einstein's proposal led Robert A. Millikan, of the Ryerson

Laboratory of the University of Chicago, to carry out an exhaustive experimental study. Millikan measured the kinetic energy of photoelectrons as a function of the frequency of the light that produced them. He found that Einstein's equation, Eq. (7.1), was an excellent fit to the facts, and concluded that the concept of a quantum of radiation must be presumed to be a valid one.

Millikan's results were first reported to a meeting of The American Physical Society in April, 1914; and a detailed description was published in 1916. The status of the new theory is well described in the following quotation from the introduction to that paper:

> It was in 1905 that Einstein made the first coupling of photo effects and with [*sic*] any form of quantum theory by bringing forward the bold, not to say the reckless, hypothesis of an electromagnetic light corpuscle of energy $h\nu$, which energy was transferred upon absorption to an electron. This hypothesis may well be called reckless first because an electromagnetic disturbance which remains localized in space seems a violation of the very conception of an electromagnetic disturbance,[1] and second because it flies in the face of the thoroughly established facts of interference. The hypothesis was apparently made solely because it furnished a ready explanation of one of the most remarkable facts brought to light by recent investigations, *viz.*, that the energy with which an electron is thrown out of a metal by ultraviolet light or x-rays is independent of the intensity of the light while it depends on its frequency. This fact alone seems to demand some modification of classical theory or, at any rate, it has not yet been interpreted satisfactorily in terms of classical theory.
>
> While this was the main if not the only basis of Einstein's assumption, this assumption enabled him at once to predict that the maximum energy of emission of corpuscles under the influence of light would be governed by the equation
>
> $$\tfrac{1}{2}mv^2 = V \cdot e = h\nu - p, \qquad \text{(I)}$$
>
> in which $h\nu$ is the energy absorbed by the electron from the light wave, which according to Planck contained just the energy $h\nu$, p is the work necessary to get the electron out of the metal and $\frac{1}{2}mv^2$ is the energy with which it leaves the surface, an energy evidently measured by the product of its charge e by the P. D. [potential difference] against which it is just able to drive itself before being brought to rest.
>
> At the time at which it was made this prediction was as bold as the hypothesis which suggested it, for at that time there were available

no experiments whatever for determining anything about how P. D. varies with ν, or whether the hypothetical h of equation (I) was anything more than a number of the same general magnitude as Planck's h. Nevertheless, the following results seem to show that at least five of the experimentally verifiable relationships which are actually contained in equation (I) are rigorously correct. These relationships are embodied in the following assertions:

1. That there exists for each exciting frequency ν, above a certain critical value, a definitely determinable maximum velocity of emission of corpuscles.
2. That there is a linear relation between V and ν.
3. That $dV/d\nu$ or the slope of the $V\nu$ line is numerically equal to h/e.
4. That at the critical frequency ν_0 at which $V = 0$, $p = h\nu_0$, i.e., that the intercept of the $V\nu$ line on the ν axis is the lowest frequency at which the metal in question can be photo-electrically active.
5. That the contact E. M. F.[2] between any two conductors is given by the equation

$$\text{Contact E. M. F.} = h/e\ (\nu_0 - \nu_0') - (V_0 - V_0').$$

No one of these points except the first had been tested even roughly when Einstein made his prediction and the correctness of this one has recently been vigorously denied by Ramsauer. As regards the fourth Elster and Geitel had indeed concluded as early as 1891, from a study of the alkali metals, that the more electro-positive the metal the smaller is the value of ν at which it becomes photo-sensitive, a conclusion however which later researches on the non-alkaline metals seemed for years to contradict.

During the ten years which have elapsed since Einstein set up his equation the fifth of the above assertions has never been tested at all, while the third and fourth have never been subjected to careful experimental test under conditions which were even claimed to permit of an exact and definite answer, nor indeed can they be so subjected without simultaneous measurements in vacuo of both contact potentials and photo-potentials[3] in the case of metals which are sensitive throughout a long range of observable frequencies. In making this statement I am not under-rating at all the exceptionally fine work of Richardson and Compton, who in common with most other observers interpreted their results in terms of Einstein's equation, but who saw the significance of that equation much more clearly than most of their predecessors had done. I am merely calling attention to the fact that the slope mentioned in (3) and the intercept mentioned in (4) cannot possibly be determined with any approach to certainty unless the region of wave-lengths open to study is larger than it is in the case of any save the alkali metals, and also, in the case of (4) unless simultaneous measurements are made in vacuo upon photo-potentials and contact E. M. F.'s.

As the last paragraph of the foregoing quotation implies, there had been earlier attempts at verification of Einstein's proposal. In particular, very careful studies had been reported in 1912 by A. Ll. Hughes from Cambridge, England, and, as mentioned, by O. W. Richardson and A. H. Compton from Princeton. Nevertheless, a review of the subject published in 1913 concluded that the case was still open.

The work at the University of Chicago had begun in 1905, presumably immediately upon publication of Einstein's paper, and various aspects of the work had been reported in the years from 1907 to 1912. Gradually, these earlier results, as Millikan puts it,

> . . . revealed the necessity of questioning the validity of all results on photopotentials unless the effects of surface films are eliminated either by removal of the films or by simultaneous measurement in vacuo of photopotentials and contact E. M. F.'s or by both procedures at once. Accordingly I have initiated in 1910 on a somewhat elaborate scale simultaneous measurements on photoeffects and contact E. M. F.'s in vacuo on film free surfaces.

The investigation required a number of more or less elaborate precautions, some of which will be discussed later. The most bothersome, however, was the one mentioned in the last quotation, that of eliminating the effects of surface films. The obvious solution is to prepare the surfaces under such conditions that the films are prevented from forming; and, despite the formidable mechanical difficulties involved, this was the course taken.

> In all of this photoelectric work at the Ryerson Laboratory the same general method has been employed, namely, the substances to be studied or manipulated have been placed in the best obtainable vacuum on an electromagnetically controlled wheel and all the needed operations have been performed by movable electromagnets placed outside. As new operations have been called for the tubes have by degrees become more and more complicated until it has become not inappropriate to describe the present experimental arrangement as a machine shop in vacuo. The operations which are now needed in all the tubes which are being used are:
>
> 1. The removal in vacuo of all surface films from all surfaces.

2. The measurement of the photocurrents and photopotentials due to these film-free surfaces.

3. The simultaneous measurement of the contact E. M. F.'s of the surfaces.

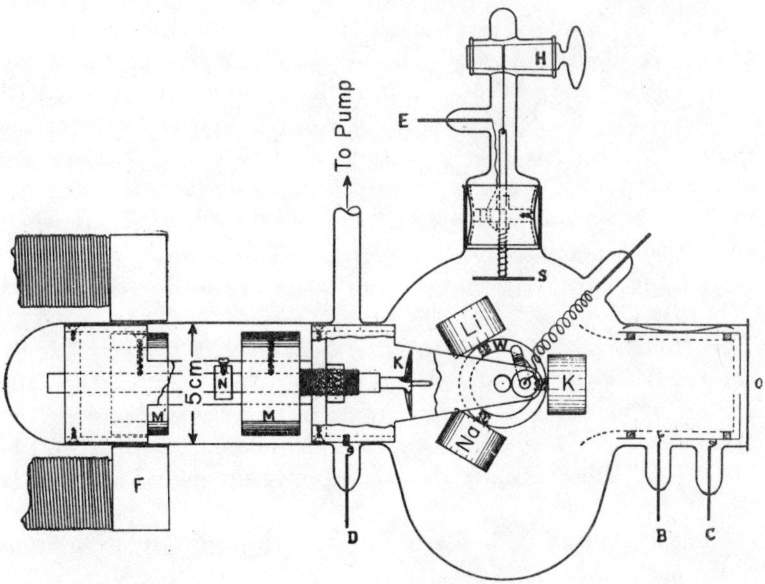

FIG. 7.1 Diagram of one of the tubes used by Millikan in his study of the photoelectric effect. [*Phys. Rev. 7*, p. 362 (1916), Fig. 2; some lettering replaced for legibility.]

A diagram of one of the tubes appears in Fig. 7.1. The three cylinders carried on the wheel *W* are cast from the metals to be studied. The wheel itself can be rotated by an electromagnet, not shown, so as to bring any of the cylinders opposite any of the other parts of the apparatus. *K* is a rotating knife, which can be moved back and forth along the axis, and rotated, by the action of the electromagnet *F* on the armatures *M* and *M′*. The metal to be studied is first brought opposite the knife, which is then advanced far enough to cut a thin shaving off the face of the cylinder; the cut is made by rotating the knife, which is then retracted again, the shaving falling down below the wheel where it helps to remove any residual oxygen from the bulb. The fresh surface can then be turned to face either the plate *S*, for measurement of the contact emf, or the window *O* and electrodes *B* and *C*, for measurement of photoelectric behavior.

Electrodes *S* and *B* were made of copper, carefully treated so as to have identical contact potentials. The measurement of the contact emf was based on the fact that if there were a potential difference, of whatever origin, between *S* and the cylinder, a change in the separation of *S* and the cylinder (by winding the thread supporting *S* up on the stopcock) would cause some charge to move through an electrometer connected to them. Thus, when an external potential was applied so as to give *no* motion of charge, the external potential must be just canceling the contact potential.

In studies of the photocurrents themselves, a beam of monochromatic light entered through the window *O* and struck the surface being studied. The photoelectrons were collected by the double cylinder *B* and *C*, which were insulated from each other inside the tube but, in the actual measurements, electrically connected outside.

One precaution is discussed in the following quotation from the paper:

> Since the aim was to test with the utmost possible accuracy the slope of the line connecting frequency with the maximum P. D. it was necessary first to know ν with great precision and second to see that no trace of light of frequency greater than that being plotted[4] got through the slit of the spectroscope. To this end a . . . mercury lamp was used as a source . . . and only such lines were chosen for use as had no companions anywhere on the short wave-length side. . . . Light filters to cut out stray short wave-length light were also used. . . . Since the measurement was to be made on the maximum P. D. and since this increases with decreasing wave-length it was not of great importance that the source be of great purity on the long wave-length side.

Millikan also considered and, as far as possible, eliminated errors which had plagued other workers. One was "back leakage" of photoelectrons liberated from the collecting electrode by reflected light. If such electrons are present, the apparent critical retarding potential will actually be the potential at which the "forward" current just balances the back leakage. Richardson and Compton had been particularly conscious of this problem. Millikan avoided it, for all but one of the wavelengths he used, by using a collecting electrode whose own photoelectric threshold wavelength was shorter than that of the incident light; he corrected for it in the one remaining case. Another source of error was the fact that previous investigators had used a very small range of wavelengths, none extending over a range equal to the smaller limit, so that the workers were forced to try to deduce the shape of a curve from a very short portion of it. Millikan used a range nearly four times the lower limit. A third source of error was light of short wavelength which reached the sensitive surface by diffuse reflection in the monochromator. This problem was eliminated, in cases where it was critical, by the use of light filters; usually, however, they were not necessary.

The currents in the critical range were very small, so they were measured (as was common at the time) as charge transported in a given time—in this case, 30 seconds—by means of a quadrant electrometer.[5] The resulting data were plotted as curves of photocurrent versus potential difference for each of several wavelengths. A set of such curves is shown in Fig. 7.2. The intercept of each curve with the horizontal axis gives the value of *V* for that wavelength.

> It will be seen . . . that the maximum possible error in locating any of the intercepts is say two hundredths of a volt and that the total range of volts covered by the intercepts is more than 2.5. Each point, therefore, of a potential-frequency curve should be located with not more than a per cent of uncertainty. The frequencies are, of course, known with great precision.

The values of the intercepts were then plotted versus wavelength. The plot obtained from the curves of Fig. 7.2 is shown in Fig. 7.3. "It will be seen that the first result is to strikingly confirm the conclusion . . . as to the correctness of the pre-

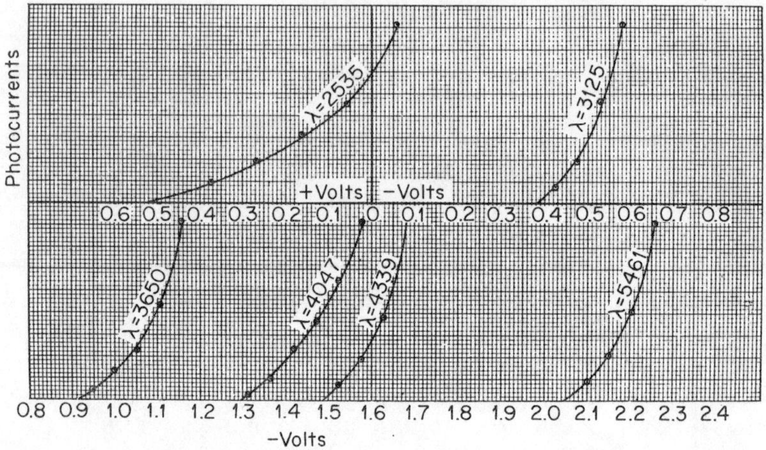

FIG. 7.2 A set of photocurrent curves for sodium. These curves were used to provide data for the curve in Fig. 7.3. [*Phys. Rev. 7*, p. 371 (1916), Fig. 5; lettering replaced for legibility.]

dicted linear relationship between maximum P. D. and ν, no point missing the line by more than about a hundredth of a volt."

If Eq. (7.1) is divided by e, the result is

$$V = (h/e)\nu - (h/e)\nu_0,$$

a straight line whose slope is h/e. Thus, from the measured slope of the line in Fig. 7.3 and Millikan's own earlier determination of e, he could compute h. The result was 6.56×10^{-27}, in complete agreement with the value originally computed by Planck from the constants of the blackbody radiation laws. Moreover, many determinations, on different surfaces, gave the same results to within experimental uncertainty.

Critical frequencies were determined in two ways; the agreement between them would be a test of the fourth point in Millikan's list. It will be recalled that the experimental apparatus was so devised that Millikan could measure the contact emf between the photosensitive surface and a surface identical to that of the collecting electrode. The actual potential difference between the emitting and collecting electrodes was equal to the

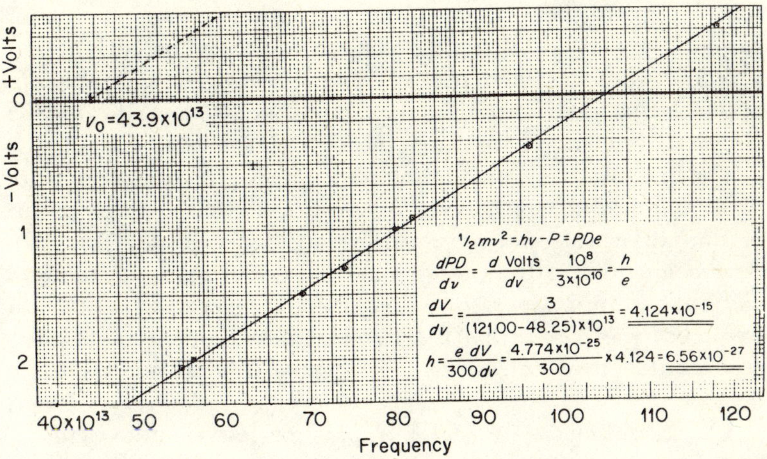

FIG. 7.3 The plot of maximum energy, measured as stopping potential difference, versus frequency for sodium. The inset outlines the computation of Planck's constant *h* from the measured slope of the line. The dashed curve is discussed in the text. [*Phys. Rev. 7*, p. 373 (1916), Fig. 5; lettering replaced for legibility.]

measured, applied voltage minus the contact potential.[6] The curve of actual potential difference versus frequency, therefore, could be obtained from the "raw" curve of Fig. 7.3 by shifting it upward by an amount equal to the contact emf. The resulting curve is shown dashed in Fig. 7.3. It is this curve whose intercept with the frequency axis gives ν_0.

The second method made use of another property of the photocurrent. If the intensity and frequency of the stimulating light are fixed, and a variable potential difference applied so as to accelerate the electrons toward the collector rather than to stop them, the collected current at first increases with increasing voltage but eventually reaches a limiting value, the saturation current. For fixed frequency, the saturation current is directly proportional to the intensity. The saturation per unit intensity, therefore, is some sort of function of the frequency; the exact nature of the function depends on the nature of the surface at least to the extent of being zero for frequencies less than the critical frequency. The procedure, then, was to measure the intensity of the incident radiation, by means of a thermopile,[7] together with the corresponding saturation photocurrent, for each of several wavelengths. The saturation photocurrent per unit intensity was then plotted against wavelength. The point at which the curve crosses the wavelength axis is the critical wavelength λ_0, related to ν_0 by $\lambda_0\nu_0 = c$.

The use of this method involved a considerable amount of luck, as nothing was known in advance as to the likely shape of the curve. It could have turned out to be a step function, zero below the critical frequency and some constant nonzero value above, or at any rate to have such a sharp rise as the critical frequency was passed that the accurate measurement of its intercept would prove unfeasible. Fortunately, this proved not to be the case.

The second method was also, in practice, less accurate than the first. In the examples shown in Millikan's paper, each curve was determined by only three or four points and then extrapolated. Millikan admits, "I should be unwilling to claim that the direct observations fix the long-wave-length limit with a precision greater than 100 Å. Nevertheless," he concludes, "these observations seem to leave no room whatever for doubt that the agreement demanded by Einstein's equation between the two methods of determining ν_0 actually exists."

The fifth point mentioned in Millikan's introduction need not concern us further, except to mention that it was satisfactorily validated.

Millikan concludes his paper with five pages of discussion of theories of photoemission. Most of this material has been made obsolete by later developments. Two portions, however, remain valid and warrant quotation:

> Perhaps it is still too early to assert with absolute confidence the general and exact validity of the Einstein equation. Nevertheless, it must be admitted that the present experiments constitute very much better justification for such an assertion than has heretofore been found, and if that equation be of general validity, then it must certainly be regarded as one of the most fundamental and far reaching of the equations of physics; for it must govern the transformation of all short-wave-length electromagnetic energy into heat energy. . . .

The photoelectric effect then, however it is interpreted, if only it is correctly described by Einstein's equation, furnishes a proof which is quite independent of the facts of black-body radiation of the correctness of the fundamental assumption of the quantum theory, namely, the assumption of a discontinuous or explosive emission of the energy absorbed by the electronic constituents of atoms from other waves. It materializes, so to speak, the quantity "h" discovered by Planck through the study of black body radiation and gives us a confidence inspired by no other type of phenomenon that the primary physical conception underlying Planck's work corresponds to reality.

FOOTNOTES

[1] It must be remembered that the whole concept of an electromagnetic wave implies that an electromagnetic disturbance which is initially localized will spread in all directions, just as ripples do on the surface of water.

[2] When two different metals are electrically connected, either by direct contact or through an electrical circuit, a difference in electrical potential exists between them *in addition to* any potential difference that may be externally applied. This difference is the *contact potential difference* or *contact electromotive force* (*emf*). It does not affect the flow of current in the circuit, as the sum of all the contact emf's around a complete circuit vanishes. It does, however, contribute to static effects, and particularly to the actual potential difference between two electrodes in an electron tube; this is the basis for the comment in the final sentence of the quotation, as well as for one of the methods described later for determining critical frequencies.

[3] That is, potentials required to suppress photoelectric currents.

[4] The reader should recall that, as stated in the second paragraph of this chapter, it was already known that an increase in frequency produced an increase in the energy of the emitted electrons.

[5] See Appendix A.

[6] See footnote 2, p. xx.

[7] A thermopile is a group of thermocouples, connected electrically so as to act in combination and blackened so as to absorb incident radiation.

BIBLIOGRAPHY

The review article mentioned on page 80 is R. Pohl and P. Pringsheim, *Philosophical Magazine* **26,** 1017–1024 (1913).

Millikan's own work is published as R. A. Millikan, *The Physical Review* **7,** 355–388 (1916).

8 *Spatial Orientation of Atomic Magnets*

Bohr's theory of atomic structure was published in a series of papers between 1913 and 1915,[1] and by the early 1920's had been fairly thoroughly developed by many workers, notably Arnold Sommerfeld. One of its fundamental postulates was that certain dynamical quantities relating to periodic motion could not take on arbitrary values, but only a set of discrete values, integral multiples of Planck's constant h. This postulate accounted nicely for the discrete energy states deduced from the experiment of Franck and Hertz described in Chapter 6—not only for their qualitative existence, but, in the case of the hydrogen atom, for their quantitative values as well.

An even more curious result of the application of this rule is the effect that became known as "spatial quantization." This does not imply the quantization of space itself, but rather of purely spatial aspects of the behavior of a dynamical system. In particular, suppose that in every atom of a particular element, the complement of electrons has an angular momentum different from zero. The rule states that when such an atom is in a magnetic field, the angular momentum vector can make only certain well-defined, discrete angles with the direction of the field. The angles permitted by the theory are those such that the component of the angular momentum along the field is an integral multiple (positive or negative, but not zero) of $h/2\pi$ where h is Planck's constant.[2] The reason that this result was regarded as curious was just the sort of extrapolation discussed in Chapter 1, for classically the value of the component of a vector along any specified direction is restricted only by the condition that it cannot exceed the magnitude of the vector.

The feature that makes this theory susceptible to test is that such an atom will also possess the properties of a magnetic dipole, with dipole moment along the direction of the total angular momentum. A magnetic dipole in a nonuniform magnetic field experiences not only a torque tending to align it with the local direction of the field, but also a force whose magnitude depends on the angle between the dipole and the gradient of the magnetic field. Consequently, a beam of such atoms passing through a nonuniform magnetic field will be deflected. If the orientation of the atomic magnets relative to the field is unrestricted, as the classical ideas indicated, the deflections of the individual atoms will have values throughout some range, and the beam will be spread. If, on the other hand, the atomic magnets can lie only along certain directions, as predicted by quantum theory, then the deflections will likewise be restricted to a few values, and the beam will be split into several parts. This was what Otto Stern had in mind when he wrote, in a paper published in 1921, "Now, whether the quantum theoretical or the classical conception is valid can be distinguished by means of an experiment, completely simple in principle. One needs only to study the deflection that a beam of atoms undergoes in a suitably inhomogeneous magnetic field."

The quantitative argument runs as follows: An electron moving in one of the orbits of Bohr's theory, with an angular momentum $\vec{L}$, has a magnetic moment $\vec{\mu}$ which is proportional to $\vec{L}$ and along the same line. (In an atom containing several electrons, both the angular momenta and the magnetic moments of the individual electrons add vectorially, so that the total magnetic moment is related in this same way to the total angular momentum.) When the atom is in a magnetic field $\vec{B}$, the action of the field on the magnetic moment produces a torque in a direction perpendicular to both $\vec{\mu}$ (and therefore $\vec{L}$) and $\vec{B}$, and of magnitude proportional to

B and to L, and dependent on the angle between $\vec{B}$ and $\vec{L}$. The effect, according to classical dynamics, is to cause $\vec{L}$ to precess about $\vec{B}$—to trace out a cone whose axis is along $\vec{B}$. The frequency of precession is proportional to B, the factor of proportionality involving only the properties of the electron and natural constants; for a field of 10^3 gauss, the period is 7×10^{-10} second. Now suppose that the field is not uniform, but varies from point to point. Denote by $\partial\vec{B}/\partial x$ the change in $\vec{B}$ per unit displacement in the x direction, and similarly for y and z. Then there will be a net force $\vec{F}$ on the atom, given by

$$\vec{F} = \mu_x \partial\vec{B}/\partial x + \mu_y \partial\vec{B}/\partial y + \mu_z \partial\vec{B}/\partial z. \tag{8.1}$$

Because of the precession, the components of $\vec{\mu}$ perpendicular to $\vec{B}$ vary sinusoidally; a common practice is to choose the z axis along $\vec{B}$, so that the varying components are μ_x and μ_y. If Eq. (8.1) is averaged over a time long compared with the period of precession, this sinusoidal variation makes the first two terms average to zero, and there remains only

$$\langle\vec{F}\rangle_{\text{av}} = \mu_z \partial\vec{B}/\partial z.$$

The atom, therefore, is accelerated in the direction of $\partial\vec{B}/\partial z$.

Now imagine an electromagnet whose poles are shaped as shown in Fig. 8.1; the z axis is vertical. Close to the knife-edge and just above it, the field $\vec{B}$ is strongest and the direction of $\partial\vec{B}/\partial z$ is along $\vec{B}$. An atom of mass M traveling along parallel to the knife-edge and just above it will experience an acceleration $a = \langle F\rangle_{\text{av}}/M = \mu_z \, |\, \partial\vec{B}/\partial z \,|\, /M$, in a direction along $\partial\vec{B}/\partial z$ and thus perpendicular to its initial motion. If it spends a time t traversing the field, it will have been deflected from its original path by an amount $z = \frac{1}{2}at^2 = \frac{1}{2}\mu_z \, |\, \partial\vec{B}/\partial z \,|\, t^2/M$. The

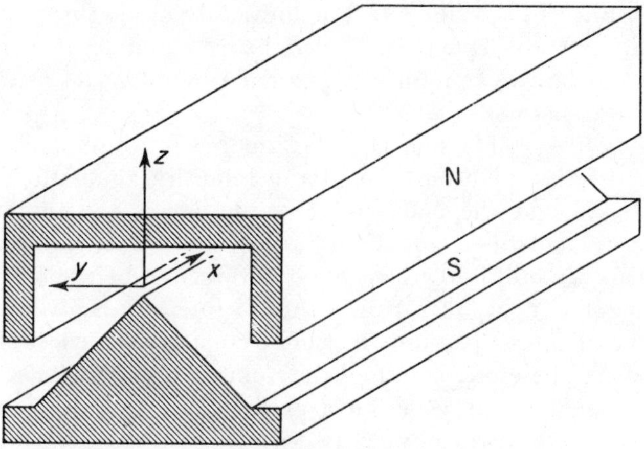

FIG. 8.1 Shape of magnet poles to give, in the region just above the knife-edge, a large value of $\partial\vec{B}/\partial z$ in the direction of $\vec{B}$.

time t is the speed v at which the atom travels divided into the length l of the magnet poles. The deflection, then, is

$$z = \frac{1}{2}\,\frac{\mu_z}{M}\left|\frac{\partial\vec{B}}{\partial z}\right|\frac{l^2}{v^2}. \tag{8.2}$$

Now consider an atom for which $L = h/2\pi$. Then according to the 1921 version of quantum theory, the component L_z of $\vec{L}$ along $\vec{B}$ can only be $\pm h/2\pi$, and μ_z can be only $\pm\mu$. An initial beam of atoms all having the same velocity would then be split into two by the action of the field, and there would be no part of the beam undeflected.

The classical case is a bit more complicated. Let us write Eq. (8.2) as

$$z = C\mu_z$$

or

$$z = C\mu\cos\theta, \tag{8.3}$$

where θ is the angle between $\vec{\mu}$ and $\vec{B}$. The classical concept is that θ can have all possible values, so that $\cos\theta$ will range continuously from -1 to $+1$. The initial beam is not split but spread, with the extent of spread being equal to the amount by which the two components predicted by quantum mechanics would be separated.

It is still possible that the classical result would look somewhat like the quantum one, if the intensity along the spread had maxima at the ends and a minimum at the center. We can show that this is not the case, provided we assume that the moments are uniformly distributed in direction before they enter the magnetic field. The distribution of intensity is given by the number of atoms per unit deflection at a given deflection; the procedure, therefore, is to find the number of atoms for which θ is in a small range $d\theta$ about a given value, and the values of the deflection corresponding to this same range, and take the ratio of the former to the latter. It is shown in calculus that if z and θ are related as in Eq. (8.3), with θ in radians, then the range dz of z corresponding to a very small range $d\theta$ of θ is given by

$$dz = C\mu \sin \theta \, d\theta, \tag{8.4}$$

which is the denominator of our ratio. The numerator is obtained from the fact that if a large number of vectors $\vec{\mu}$ are oriented randomly, their tips are distributed uniformly over a sphere of radius μ; the number whose angle with a specified direction lies in the range $d\theta$ around θ is the number whose tips lie in a band of width $\mu \, d\theta$ centered at colatitude[3] θ (see Fig. 8.2), and is therefore proportional to the area of such a band. That area is the circumference of the band, $2\pi a$, times its width, $\mu \, d\theta$, provided $d\theta$ is small. The radius a is equal to $\sin\theta$ times the radius μ of the sphere, so that the area of the band is $2\pi\mu^2 \sin\theta d\theta$. Thus, the number of atoms whose magnetic moments make an angle with $\vec{B}$ in the range $d\theta$ around θ is proportional to $\sin\theta d\theta$. When this is divided by Eq. (8.4), the angular dependence cancels out. The result is that according to classical theory, the initial spot is spread into a band of uniform intensity.

These calculations, however, have assumed a beam involving a single velocity. In actuality, the beam is obtained by vaporization and contains a distribution of velocities. This turns out to improve matters. In the quantum case, each of the two spots is spread out somewhat, but the central region is still a *minimum* [Eq. (8.2) shows that if μ_z cannot be zero, and $\partial\vec{B}/\partial z \neq 0$, a zero deflection can result only from an infinite velocity.][4] In the classical case, on the other hand, the single uniform streak is replaced by a superposition of streaks whose lengths range from very small to very large; the combination produces a *maximum* of intensity at the center. It is evident that the distinction is, as Stern maintained, clear cut.

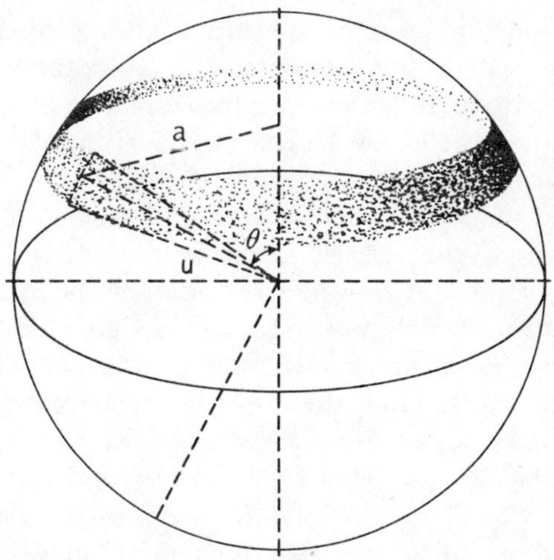

FIG. 8.2 Band at colatitude θ on a sphere of radius μ.

However, while the experiment appears simple in principle, it turns out in practice to be another matter. To begin with, the experiment must be carried out in vacuum so that the beam will not be destroyed by scattering from gas molecules. Considered in conjunction with the state of vacuum techniques at the time, this restricted the possible length l of the poles to a few centimeters. It is possible to make B of the order of 10^3 gauss, and $|\partial\vec{B}/\partial z|$ of the order of 10^4 gauss per centimeter. The value of μ is known in terms of fundamental constants.

The combination Mv^2 is twice the kinetic energy of the atom; the kinetic energy, in turn, is determined by the vapor temperature, which may be of the order of 1000°K. When such numbers are put into the formula, the deflection that can be expected is found to be of the order of 0.01 mm. The experiment was possible, but it would evidently be a delicate one.

When Stern submitted the paper containing the foregoing analysis, in late August of 1921, he and a co-worker, Walther Gerlach, were already occupied with carrying out the experiment at Frankfurt am Main, Germany; and by the middle of November they had preliminary results—too preliminary, however, to permit a decision on the main question. They proceeded to make some improvements on the apparatus (a footnote to the third paper comments that "It was possible for these to be worked out and tested by joint efforts during the Christmas vacation"), and submitted firm conclusions on 1 March 1922.

Compared with present-day ultrahigh-precision technology and miniaturization, the experimental arrangement is not exceptional; but at the time it was a considerable feat. The substance used was silver, which was vaporized in an electrically heated oven and escaped through a circular opening of area 1 mm^2. At a distance of 2.5 cm from the oven[5] was a diaphragm containing an approximately circular aperture of area 3×10^{-3} mm^2, i.e., radius about 0.03 mm. Another 3.3 cm beyond this was a second diaphragm, whose opening was in the form of a slit of length 0.8 mm and width 0.03 to 0.04 mm, oriented perpendicular to the direction of B. Such tiny dimensions were obviously necessary to produce a beam which would not be vastly larger than the amount by which it was deflected. The slit was placed just at the apex of one end of the knife-edge pole piece (see Fig. 8.1), and the set of openings was so adjusted that the beam traveled parallel to the knife-edge. The magnet poles were 3.5 cm long. The whole arrangement was housed in a casing whose walls were thick enough so that pressure on the magnet poles, resulting from the vacuum inside the chamber, would not shift the relative position of the components. The "exposure times" were eight hours; even then, the deposit on a glass plate at the far end of the pole pieces was too thin to be visible, and had to be darkened by precipitation of nascent silver.

The best exposure is reproduced in Fig. 8.3. Gerlach and Stern say:

> Two other exposures yielded a result identical in all essential points, not, however, with this complete symmetry. It must be said here that a reliable adjustment of such small diaphragms by optical means is very difficult, so that the achievement of such a completely symmetric exposure as in Fig. 3 [reproduced as Fig. 8.3] is surely in part the result of luck; misplacement of one diaphragm by a few hundredths of a millimeter is already enough to make an exposure fail.

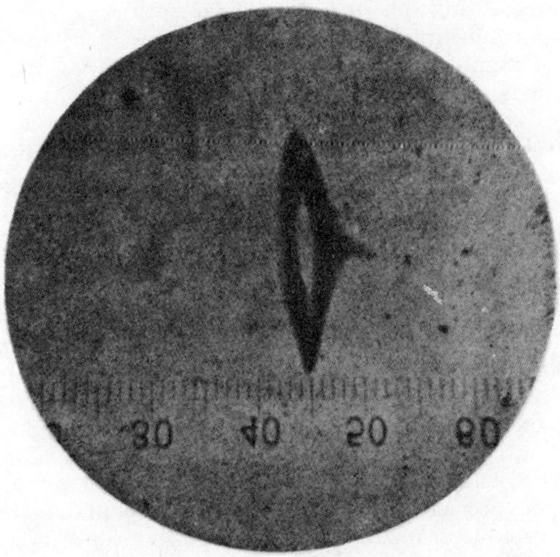

FIG. 8.3 Splitting of a beam of silver atoms by an inhomogeneous field. Each scale division is 1/20 mm. [Z. *Phys.* 9, p. 350 (1922), Fig. 3. The reversal is as in the original.]

The one characteristic common to all three exposures was the clear separation of the beam into two components. As the authors put it, "The splitting of the atomic beam in the magnetic field gives rise to two discrete beams. There are no undeflected atoms detectable." [6]

Reference to the preceding discussion shows that this result clearly confirms the quantum hypothesis as opposed to the classical behavior. The terminology of the time was used by Gerlach and Stern: "We behold in these experimental results the direct experimental proof of the quantization of direction in a magnetic field."

FOOTNOTES

[1] James Franck remarked in 1961 that he and Hertz had not known of Bohr's work while their own experiments (see Chap. 6) were going on, but that even if someone had told them it existed they probably would not have read it: The problem of atomic structure was regarded as so arcane that anyone claiming to have solved it was likely to be a crackpot.

[2] The theory also maintained that the magnitude of the angular momentum must be an integral multiple of $h/2\pi$. Subsequent developments have revealed that half-integral multiples of $h/2\pi$ are also allowed for the magnitude, and that when the magnitude is an integral multiple of $h/2\pi$, the component can be zero. If the latter feature had been known in 1921, the experiment described in this chapter would not have been regarded as a stringent test of the theory, as will be made clear; fortunately, the experimenters happened to work with an element which falls into the half-integral class, for which the zero component is excluded.

[3] The colatitude of a point on a sphere is the angle that a radius to that point makes with the axis of the sphere. It is 90° minus the latitude.

[4] Note here the importance of the theoretical error according to which μ_z supposedly could not be zero. If the possibility that $\mu_z = 0$ had been allowed for in the quantum case, the distinction between the two cases would be that of three maxima with intervening minima in the quantum case compared with a single central maximum in the classical case. Considering the difficulty and the unknown sensitivity of the experiment, Stern might well have concluded that it was not feasible.

[5] This distance had been only 1 cm in the preliminary work, but was increased so as to prevent the aperture from being plastered over either by molten silver splattered out of the oven or by too rapid incrustation by deposition from the beam.

[6] In the light of subsequent developments (cf. footnote 2), this is found to be a fortuitous result. If Stern and Gerlach had used sulfur, for example, instead of silver, the beam would have been split into five components instead of two, and one of the five would have undergone no deflection. It is intriguing to speculate on how such a result would have been interpreted.

BIBLIOGRAPHY

All three of the papers discussed in this chapter are in German. They are O. Stern, *Zeitschrift für Physik* **7**, 249–253 (1921); O. Stern and W. Gerlach, *ibid.* **8**, 110–111 (1922), and **9**, 349–355 (1922). A translation of the last is given in *The World of the Atom,* Vol. 2, pp. 936–939.

9 *Particle Properties of Light*

X rays were identified as electromagnetic waves in 1912, by means of diffraction experiments suggested by von Laue and carried out by Friedrich and Knipping. The behavior of electromagnetic waves was well understood on the basis of Maxwell's theory. In particular, it was an easy matter to envisage a mechanism by which they were scattered, and it was a straightforward procedure to compute the quantitative features of the scattering. The mechanism was that the varying electric field of the wave would set the electrons in the scatterer into forced vibration; the electrons, in turn, because they were being accelerated, would emit radiation. The scattered radiation would be of the same frequency as the incident—it had to be, as it was just the frequency of vibration of the emitting electrons. It would have all the properties of the radiation emitted by an oscillating electric dipole: intensity symmetrically distributed around the line of the electron's motion and, in any plane containing that line of motion, varying as the square of the sine of the angle between the line of motion and the direction of propagation; and polarization properties that are of no concern in the present discussion. Moreover, the fraction of the incident energy transferred into the scattered radiation should be independent of the frequency.

In the ten years following 1912, this theory met with steadily increasing difficulty. The first discrepancy was that for x rays of very short wavelength, or for γ rays, the scattered intensity was greater in the forward direction (that is, the same direction as the incident radiation) than in the backward. This feature received, for a time, a quantitative explanation by ascribing to the electron a size comparable to the x-ray wavelength and assuming that there was interference between the rays scattered by different parts of the electron. As data accumulated, however, it was found that the value that must be assigned for the diameter of the electron varied with the wavelength of the incident radiation—obviously a most unsatisfactory situation. In addition, a still more serious difficulty had appeared. It was discovered that the scattered radiation was different in frequency from the incident. Arthur H. Compton, in a paper published in 1923, had this to say about the situation:

> Such a change in wave-length is directly counter to Thomson's theory of scattering, for this demands that the scattering electrons, radiating as they do because of their forced vibrations when traversed by a primary X-ray, shall give rise to radiation of exactly the same frequency as that of the radiation falling upon them. Nor does any modification of the theory such as the hypothesis of a large electron suggest a way out of the difficulty. This failure makes it appear improbable that a satisfactory explanation of the scattering of X-rays can be reached on the basis of the classical electrodynamics.

Compton proposed, instead, to apply to the description of scattering the same quantum concept that had proved so useful in treating the photoelectric effect (see Chap. 7). From this basis he deduced that several aspects of x-ray scattering would yield results measurably different from what classical theory would predict; the most significant of these is the wavelength shift, which has become known as the Compton effect. He cited data which confirmed his theoretical ideas, and, in the case of the Compton effect, he carried out the experiment himself. His work won him the Nobel prize.

Compton expressed the basic change in viewpoint, and its consequences, as follows:

> According to the classical theory, each X-ray affects every electron in the matter traversed, and the scattering observed is that due to the combined effects of all the electrons. From the point of view of the quantum theory, we may suppose that any particular quantum of X-rays is not scattered by all the electrons in the radiator, but spends all of its energy upon some particular electron. This electron will in turn scatter the ray in some definite direction, at an angle with the

incident beam. This bending of the path of the quantum of radiation results in a change in its momentum.[1] As a consequence, the scattering electron will recoil with a momentum equal to the change in momentum of the X-ray. The energy in the scattered ray will be equal to that in the incident ray minus the kinetic energy of the recoil of the scattering electron; and since the scattered ray must be a complete quantum, the frequency will be reduced in the same ratio as is the energy. Thus on the quantum theory we should expect the wave-length of the scattered X-rays to be greater than that of the incident rays.

Having established this framework, Compton proceeded to build upon it a rather impressive structure. The first step, and the one that ultimately proved to be the most important, was to derive the relationship

$$\lambda = \lambda_0 + (2h/mc) \sin^2 \tfrac{1}{2}\theta$$

between the incident wavelength λ_0, the scattered wavelength λ, and the scattering angle θ; h is Planck's constant, m the mass of the electron, and c the speed of light. The qualitative form of the relationship is reasonable: The larger the scattering angle, the greater the momentum given to the electron; therefore the greater its kinetic energy, acquired from the incident photon, and the greater the reduction in photon frequency and the increase in wavelength. The details follow simply from application of the laws of conservation of energy and momentum to the scattering event, remembering that relativistic forms must be used for the energy and momentum of the electron. One other feature of the equation should be noted. The factor $2h/mc$ has the value 0.048 Å. It is the smallness of this quantity that prevented the detection of the effect prior to the discovery and study of x rays.

The remainder of the theoretical development was somewhat shakier. Compton noted that when the wavelength shift was reexpressed in terms of frequency, it had the same form as the expression given by classical theory for the shift resulting from scattering by an electron initially *moving* in the direction of propagation, if the velocity of the electron was properly related to the incident frequency. He then *assumed* that the relative intensities scattered at different angles would also be given correctly by the classical theory applied to that special case; and on the basis of the assumption, he derived expressions for several quantities that had been or could be measured experimentally. In every case where an experimental check was available, the results agreed better with Compton's theory than with the earlier theory.

All these calculations, however, rested upon the uncertain base of the analogy described. The crucial test would obviously be the wavelength shift, which followed solely from the quantum properties of the x rays. Here, again, some of Compton's own earlier work was relevant. He had measured the absorption of γ rays, produced by a naturally radioactive substance, after scattering at various angles. From the known variation of the absorption with wavelength, he could then determine a wavelength of the scattered radiation and compare it with that given by his theory. Once again, the agreement was satisfactory.

Even more quantitative evidence was desirable, and Compton therefore carried out another experiment, the details of which were given in a second paper published six months after the first. The procedure was, simply, to make a direct spectroscopic measurement of the wavelength of x rays at selected angles. This was done by use of the fact that a mineral crystal, such as calcite, constitutes a natural, three-dimensionally periodic array, and that when a wave is reflected by such an array, interference effects will give rise to strong maxima in the reflected intensity at angles that depend on the wavelength.[2] The apparatus, designed to be hampered as little as possible by the low intensity of the scattered beam, is shown schematically in Fig. 9.1. The target of the x-ray tube, which is the source of the x rays, is at *T;* the graphite scattering block, at *R,* in line with slits 1 and 2. The position of the tube could be changed so as to vary the scattering angle from which radiation would be accepted by the spectrometer. The tube and the scattering block were placed inside a lead box to prevent stray radiation from reaching the detector. The x rays were detected by means of the ionization they produced in a suitable chamber.

Two sets of results are shown in Fig. 9.2. The two sets were taken using slits of different widths; Compton also states that in the right-hand set, "the settings were made with the greater care,

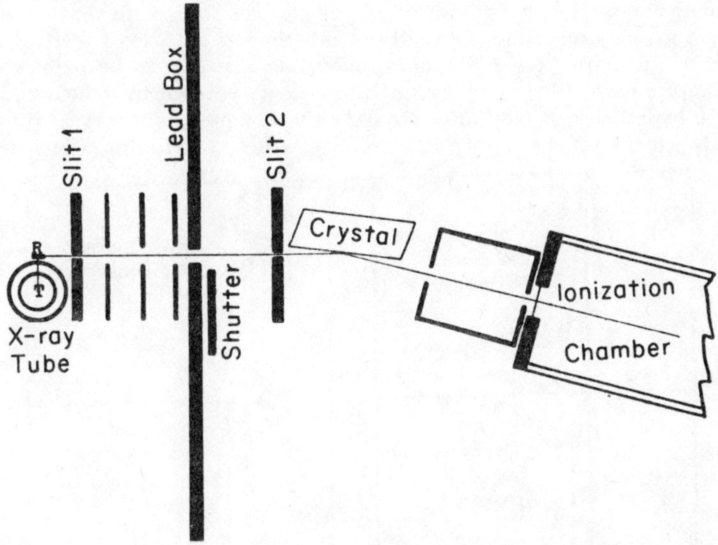

FIG. 9.1 Schematic diagram of Compton's apparatus for measuring the shift in wavelength of scattered x rays. The angle of scattering is the angle between the two segments of the beam meeting at *R*; the diagram shows the case of 90°. [*Phys. Rev.* 22, p. 410 (1923), Fig. 1; lettering replaced for legibility.]

within an experimental error of less than one minute of arc, or about 0.001 Å." He describes the results as follows:

> It is clear from these curves that when a homogeneous x-ray is scattered by graphite it is separated into two distinct parts, one of the same wave-length as the primary beam, and the other of increased wave-length. Let us call these the *modified* and the *unmodified* [*sic*[3]] rays respectively. In each curve the line *P* is drawn through the peak of the curve representing the primary line, and the line *T* is drawn at the angle at which the scattered line should appear. . . .

according to the theory he had proposed. The agreement between theory and experiment is unquestionable.[4]

In his conclusions, Compton says,

> This remarkable agreement between our formulas and the experiments can leave but little doubt that the scattering of X-rays is a quantum

phenomenon. The hypothesis of a large electron to explain these effects is accordingly superfluous, for all the experiments on X-ray scattering to which this hypothesis has been applied are now seen to be explicable from the point of view of the quantum theory without introducing any new hypotheses or constants. In addition, the present theory accounts satisfactorily for the change in wave-length due to scattering, which was

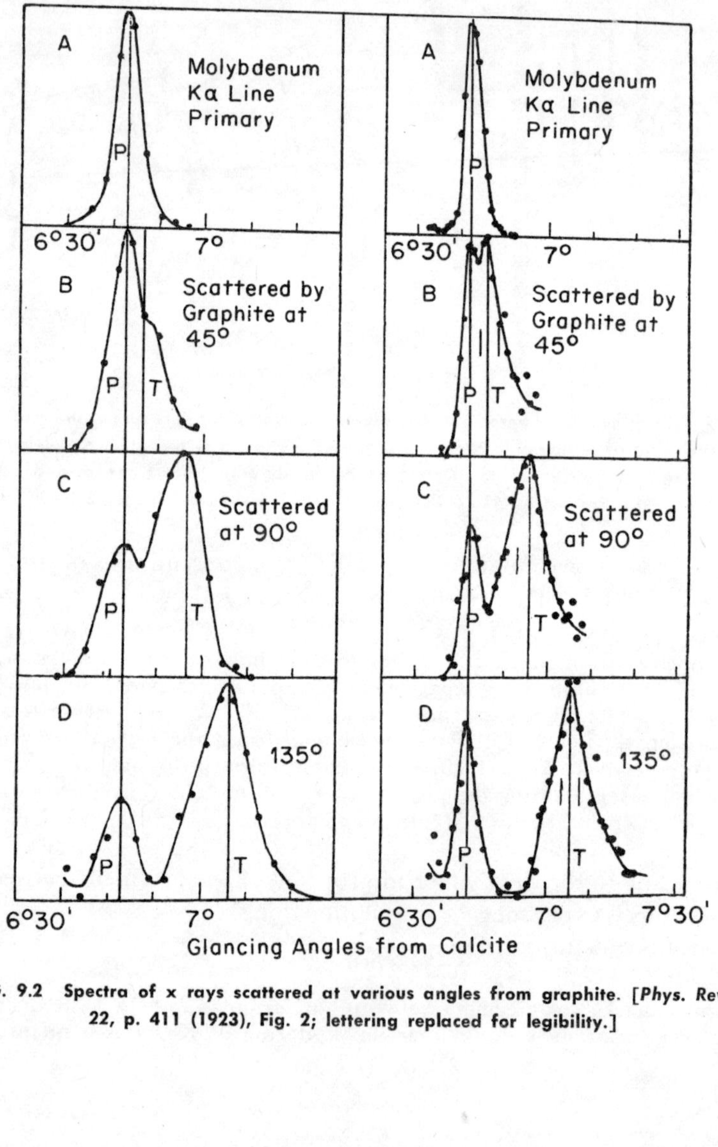

FIG. 9.2 Spectra of x rays scattered at various angles from graphite. [*Phys. Rev.* 22, p. 411 (1923), Fig. 2; lettering replaced for legibility.]

left unaccounted for on the hypothesis of the large electron. From the standpoint of the scattering of X-rays and γ-rays, therefore, there is no longer any support for the hypothesis of an electron whose diameter is comparable with the wavelength of hard X-rays.

The present theory depends essentially upon the assumption that each electron which is effective in the scattering scatters a complete quantum. It involves also the hypothesis that the quanta of radiation are received from definite directions and are scattered in definite directions. The experimental support of the theory indicates very convincingly that a radiation quantum carries with it directed momentum[5] as well as energy.

It is significant that Compton's own experiment on this effect implied something of a paradox, as Compton himself was aware. The crystal spectrometer measured a wavelike property, the wavelength, and measured it by means of a characteristically wavelike phenomenon, interference. But the effect of the graphite scatterer on the value of that wavelike property could be understood only in terms of a particlelike behavior. To Compton, "The manner in which interference occurs . . . is not yet clear. . . . In any case, the problem of scattering is so closely allied with those of reflection and interference that a study of the problem may very possibly shed some light upon the difficult question of the relation between interference and the quantum theory." This expectation was very well borne out. Within less than three years, physicists began to accept the idea that light is neither truly a wave phenomenon nor truly a stream of particles, but a separate entity whose behavior is sometimes wavelike and sometimes particlelike.

FOOTNOTES

[1] It is not clear whether Compton's assumption of momentum carried by a photon was his own or one that had been used before. The phraseology of his conclusions (see later) suggests that in either case, the idea was controversial. If it was his own contribution, he apparently expected his readers to be able to deduce for themselves that the magnitude of the momentum carried by a photon of energy E must be E/c, where c is the speed of light; certainly, there were at least two methods of deduction available to them.

[2] The effect is essentially identical to the production of a spectrum of visible light by a grating; only some details are different.

[3] This slip occurs only in the single sentence in which the terms are

originally introduced. In subsequent sentences, Compton uses the terms with their logical meanings.

[4] Further study led to the conclusion that the unshifted line resulted from scattering by electrons which were tightly enough bound in atoms that the assumptions made in the derivation of the shift were not applicable.

[5] See footnote 1.

BIBLIOGRAPHY

A. H. Compton, *The Physical Review* **21,** 483–502 (1923), and **22,** 409–413 (1923). These papers are excerpted in *The World of the Atom,* Vol. 2, pp. 911–929. Compton later gave a personal recollection of the events in A. H. Compton, *American Journal of Physics* **29,** 817–820 (1961).

10 *Wave Properties of Matter*

By the end of 1924, it had become generally accepted that the behavior of electromagnetic radiation had both some wavelike aspects and some particlelike aspects. At that point a French graduate student, Louis de Broglie, experienced a flash of genius: Why shouldn't the same be true of matter? Specifically, he suggested, the relationships between the particlelike properties energy and momentum, on the one hand, and the wavelike properties wavelength and frequency, on the other, should be the same for matter as for radiation.[1] The reason why the wave aspect had not previously been noted was the extreme smallness of the wavelengths implied for megascopic matter. A one-microgram dust mote, for example, moving at a speed of 0.1 millimeter per second,[2] would have a wavelength of only about 6×10^{-11} Å. It is necessary to go to the atomic scale of things to achieve wavelengths of the same order of magnitude as the dimensions of the object itself and thus attain any hope of observability. De Broglie's thesis examiners would not, of course, have accepted such a wild idea by itself, and so he combined it with some studies in kinetic theory. But the idea was one whose time had come, and it caught on rapidly.

Meanwhile, at Bell Telephone Laboratories, then in New York, C. J. Davisson and his co-workers were doing some fairly routine work. What happened is best told in the opening paragraphs of a paper by Davisson and L. H. Germer published [3] in 1927:

The investigation reported in this paper was begun as the result of an accident which occurred in this laboratory in April 1925. At that time we were continuing an investigation, first reported in 1921, of the distribution-in-angle of electrons scattered by a target of ordinary (polycrystalline) nickel. During the course of this work a liquid-air bottle exploded at a time when the target was at a high temperature; the experimental tube was broken, and the target heavily oxidized by the inrushing air. The oxide was eventually reduced and a layer of the target removed by vaporization, but only after prolonged heating at various high temperatures in hydrogen and in vacuum.

When the experiments were continued it was found that the distribution-in-angle of the scattered electrons had been completely changed. Specimen curves exhibiting this alteration are shown in Fig. 1 [reproduced as Fig. 10.1]. These curves are all for a bombarding potential of

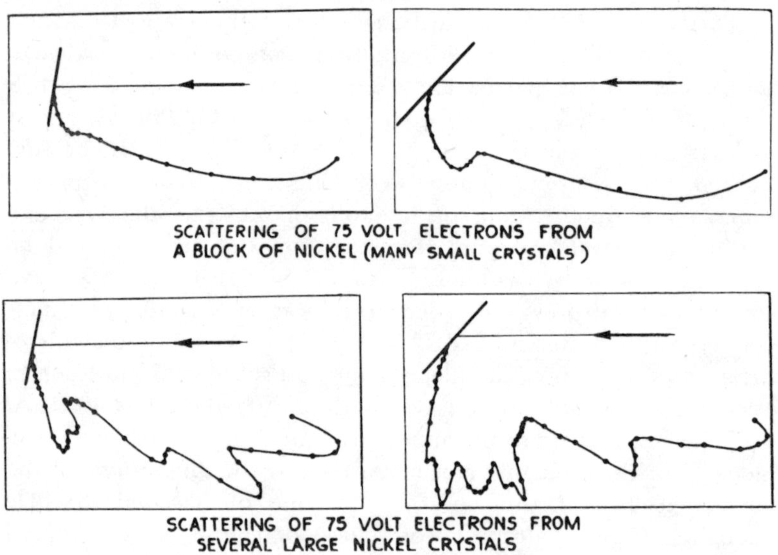

FIG. 10.1 Scattering curves from nickel before and after the accident. [*Phys. Rev.* *30*, p. 706 (1927), Fig. 1.]

75 volts. The electron beam is incident on the target from the right, and the intensities of scattering in different directions are proportional to the vectors from the point of bombardment to the curves. The upper curves (for different angles of incidence) are characteristic of the target prior to the accident. They are of the type described in the note in "Science" in 1921, and are similar to curves that have been obtained for nickel in four or five other experiments. The lower curves—obtained

after the accident—were the first of their sort to be observed. This marked alteration in the scattering pattern was traced to a re-crystallization of the target that occurred during the prolonged heating. Before the accident and in previous experiments we had been bombarding many small crystals, but in the tests subsequent to the accident we were bombarding only a few large ones. The actual number was of the order of ten.

It seemed probable from these results that the intensity of scattering from a single crystal would exhibit a marked dependence on crystal direction, and we set about at once preparing experiments for an investigation of this dependence. We must admit that the results obtained in these experiments have proved to be quite at variance with our expectations. It seemed to us likely that strong beams would be found issuing from the crystal along what may be termed its transparent directions—the directions in which the atoms in the lattice are arranged along the smallest number of lines per unit area. Strong beams are indeed found issuing from the crystal, but only when the speed of bombardment lies near one or another of a series of critical values, and then in directions quite unrelated to crystal transparency.

The most striking characteristic of these beams is a one to one correspondence, presently to be described, which the strongest of them bear to the Laue beams that would be found issuing from the same crystal if the incident beam were a beam of x-rays. Certain others appear to be analogues, not of Laue beams, but of optical diffraction beams from plane reflection gratings—the lines of these gratings being lines or rows of atoms in the surface of the crystal. Because of these similarities between the scattering of electrons by the crystal and the scattering of waves by three- and two-dimensional gratings a description of the occurrence and behavior of the electron diffraction beams in terms of the scattering of an equivalent wave radiation by the atoms of the crystal, and its subsequent interference, is not only possible, but most simple and natural. This involves the association of a wave-length with the incident electron beam, and this wave-length turns out to be in acceptable agreement with the value h/mv of the undulatory mechanics, Planck's action constant divided by the momentum of the electron.

It is worth noting that this passage points up two marks of an outstanding scientist. In the first place, when Davisson and Germer discovered something significant that they were not looking for,[4] they quickly recognized its significance. In the second place, they were ready to accept results that did not conform with expectations, and to hunt for a proper correlation of the results.

The experimental procedure was, as usual, basically simple.

Electrons from an electron gun struck a nickel target, carefully prepared so that the bombarded area was part of a single crystal. The exposed surface was a plane heavily populated with atoms, and the incident beam struck it perpendicularly. The electrons scattered in a suitable range of directions were collected, and the resulting current was measured as the experimental conditions were varied.

The essential parts of the apparatus are shown in Fig. 10.2.

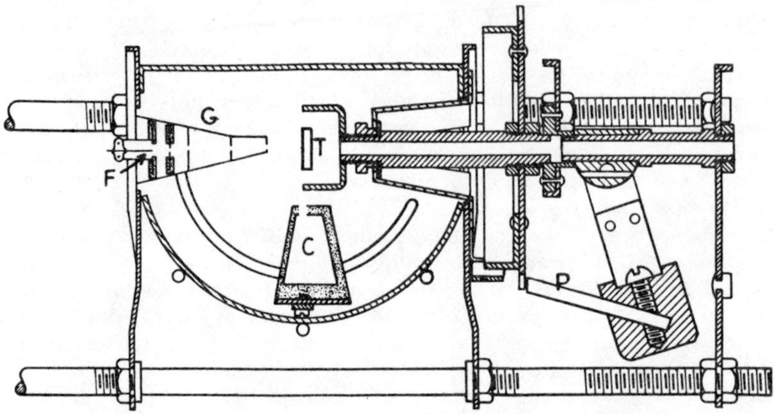

FIG. 10.2 Cross-sectional view of Davisson and Germer's apparatus. The enclosing glass bulb is not shown. [*Phys. Rev. 30*, p. 708 (1927), Fig. 2.]

The electron gun G accelerated and focused the electrons emitted by the filament F, to form a narrow beam which struck the target T. The collector C was a double-walled box, with an adjustable retarding potential applied between the outer and inner boxes so that only electrons that had undergone essentially no loss of energy reached the inner box to contribute to the measured current. The outer box, the target, and the last electrode (the outer structure and the last three diaphragms) of the electron gun were held at the same potential, so that from gun to target to collector the electrons traveled in an essentially zero electric field.

The range of directions from which electrons were accepted

was defined by the small holes in the collector boxes. The collector was suspended on an axis perpendicular to the plane of the figure and passing through the point where the incident beam struck the target; the entire apparatus could also be tilted about this axis so that the angle between the initial direction and the (central) direction of acceptance could be varied from 90° to 20°.

For reasons that will be made clear later, it was desirable also to change the relative orientation of the target and the scattering plane. This was accomplished by an ingenious arrangement. The target was mounted on a shaft to which was fixed a toothed wheel. A plunger (*P* in Fig. 10.2) was fastened to a heavy pendulum which was pivoted about an axis perpendicular to the plane of the figure. When the apparatus was tilted so that the collector viewed the target, the plunger passed between two teeth of the wheel and engaged with a milled strip of metal attached to the frame, so that the wheel and target were locked to the frame. When the tube was rotated clockwise until the main axis was slightly beyond the horizontal, the plunger disengaged from the milled strip. The suspension of the pendulum was itself free to rotate about the axis of the shaft, so that by rotating the tube about its main axis, the pendulum, plunger, wheel, and target were rotated relative to the frame and collector. This gave a range of rotation of about 20 to 30 degrees. Tilting the tube still further pulled the plunger out from between the teeth of the wheel; rotation of the tube then allowed the plunger to be moved to a different opening in the wheel. In this way the target could be rotated through the full circle. Its position was indicated by means of a scale on the wheel.

The whole arrangement, about five inches long and two inches high, was enclosed in a glass bulb and carefully evacuated, including repeated baking to liberate adsorbed and entrapped gases. The final pressure was estimated to be 10^{-8} mm of mercury.

In order to understand the results, "It is important," as Davisson and Germer put it, "to have a clear picture of the arrangement of atoms presented to the incident beam by the crystal. The nickel crystal is of the face-centered cubic type." This means that the atoms are arranged in an indefinite repetition, in every direction, of the pattern in Fig. 10.3: atoms at the corners and at the centers of the faces of a cube, which in nickel is 3.51 Å on an edge. Davisson and Germer cut, polished, and chemically etched the target so that the bombarded surface was a particular

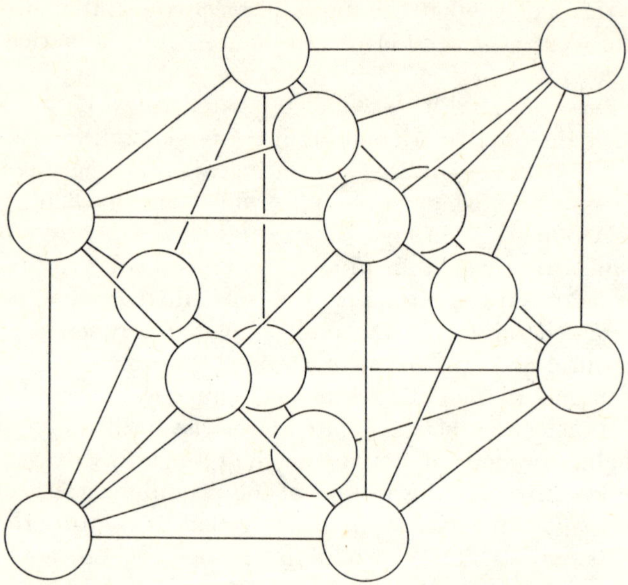

FIG. 10.3 Basic unit of the pattern of atoms in a nickel crystal.

kind of plane known as a {111} plane (for an explanation of the notation, see Appendix B). The relationship of such a plane to the basic crystal structure is shown in Fig. 10.4. The choice of this plane was made for two reasons: First, it is the plane most heavily populated with atoms; and second, it is the plane most readily developed by etching.

Their discussion continues:

> The {111}-plane is the plane of densest packing, and in this plane the atoms have a triangular arrangement. Looking directly downward onto a crystal cut to this plane (Fig. 6 [reproduced here as Fig. 10.5]) one sees the atoms of the second plane below the centers of alternate triangles of the first plane, and the atoms of the third plane below the centers of the remaining triangles.

The atoms of the fourth plane are below those of the first. The lines joining any second-layer atom with the three nearest first-layer atoms are {110}-directions in the crystal, and the lines joining it with the three next-nearest surface atoms are the orthogonal {100}-directions. It will be convenient to refer to the azimuths of these latter directions as {100}-azimuths. The azimuths of the {110}-directions are also those of the three lateral {111}-directions, . . . and we shall designate these as

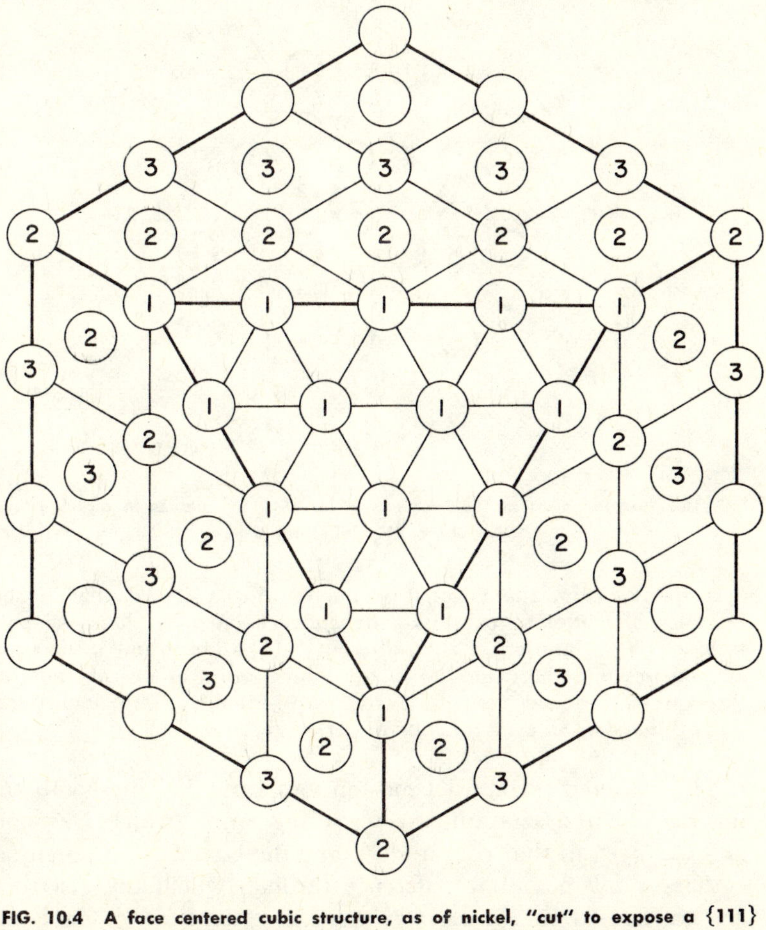

FIG. 10.4 A face centered cubic structure, as of nickel, "cut" to expose a {111} plane; the atoms in this plane are numbered "1." Also indicated are the edges of the next two successive {111} planes, numbered "2" and "3," respectively. The numbering corresponds to that used in Figure 10.5.

{111}-azimuths. We need also a designation for the azimuths that bisect the dihedral angles between adjacent members of the two sets already specified. There are six such azimuths and they will be referred to as {110}-azimuths.

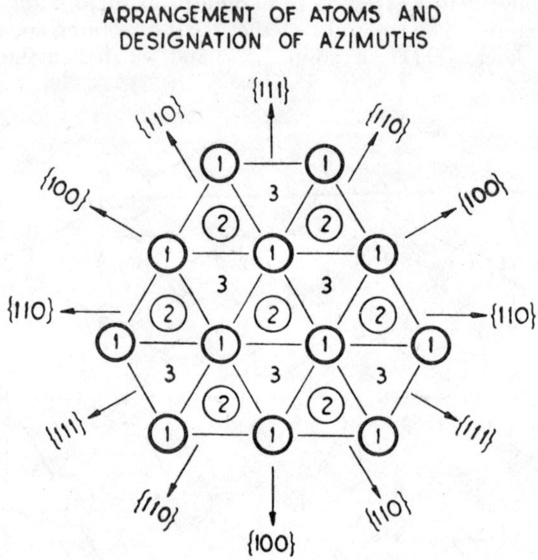

FIG. 10.5 Arrangement of the atoms in a set of {111} planes in nickel. [*Phys. Rev.* 30, p. 712 (1927), Fig. 6.]

It follows from the trigonal symmetry of the crystal that if the intensity of scattering exhibits a dependence on azimuth as we pass from a {100}-azimuth to a next adjacent {111}-azimuth (60°), the same dependence must be exhibited in the reverse order as we continue on through 60° to the next following {100}-azimuth. Dependence on azimuth must be an even function of period $2\pi/3$.

The scattered current depends on four variables: the bombarding current, the azimuth, the scattering angle (which Davisson and Germer call the "colatitude"), and the bombarding potential —that is, the potential difference through which the electrons are accelerated in the electron gun. The dependence on bombarding current is a simple proportionality and is of no further interest. There are then, so to speak, three different possible "experiments," according to which of the three other quantities was varied.

The variation with azimuth is quite easily described.

> When bombarding potential and latitude angle are fixed and exploration is made in azimuth a variation of collector current is always observed, and this exhibits always the three-fold symmetry required by the symmetry of the crystal. The curves show in general two sets of maxima—a set of three in the {111}-azimuths, and a set of three of different intensity in the {100}-azimuths. These crests and troughs are usually not pronounced.

Although exceptions to the last sentence sometimes occurred, the general form of the dependence on azimuth was much as expected. This simple, predictable behavior scarcely seems to justify the elaborate provision of means for rotating the target, but the details of the effects of other variations depended strongly on the azimuth chosen.

The truly interesting observations were obtained by fixing the azimuthal orientation at one of the three principal directions, {100}, {110}, or {111} (see Fig. 10.5), and measuring the scattered current as a function of bombarding voltage for each of a series of scattering angles. A portion of a set of curves constructed from such data is shown in Fig. 10.6.

> The general trend of a single one of these curves is not significant as it is determined in part by variation with voltage of the bombarding current.[6] The relative displacements among them, however, are significant. . . . We see, for example, that the colatitude curves for bombarding potentials near 55 volts are characterized by exceptional intensities at colatitude angles near 50°. . . . The unusual and significant feature revealed by the curves . . . is exhibited again in the set of colatitude curves on the right in Fig. 10 [reproduced here as Fig. 10.7]. We see a slight hump at 60° in the colatitude curve for 40 volts, and observe that as the bombarding potential is increased this hump develops into a strong spur which reaches a maximum development at 54 volts in colatitude 50°, then decreases in intensity and finally disappears at about 66 volts in colatitude 40°.

A similar set of spurs in the {100} azimuth attains its maximum development at 65 volts in colatitude 44°; a complete set

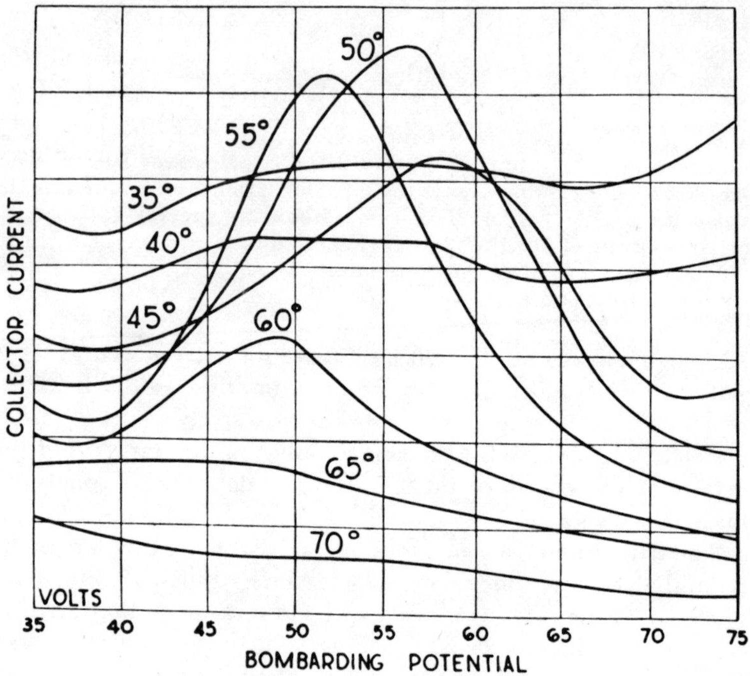

FIG. 10.6 Curves of collector current versus bombarding potential for various scattering angles. Azimuth {111}. [*Phys. Rev. 30*, p. 716 (1927), Fig. 9.]

of colatitude curves for this azimuth is shown on the left in Fig. 10.7.

This method was used in exploring the principal azimuths for bombarding potentials in the range from 15 to 350 volts. Whenever there appeared a feature of the sort just discussed, either the data were used to construct colatitude curves or they served as "a guide to voltage-colatitude ranges requiring special study." There resulted thirty sets of spurs: eleven in the {111} azimuth, twelve in the {100} azimuth, and seven in the {110} azimuth.

There are several effects that influence the position of a spur, or alter its intensity. One is the fact that because the spur is presumably "a feature superposed on the simple scattering curve the position of its maximum is falsified to some extent by the variation with angle of the background against which it appears."

This could be corrected for quite easily. Another effect was the fact that while every spur appeared in azimuth as a set of three, as demanded by symmetry, the symmetry was not perfect. This was found to be ascribable to a slight misalignment, about one degree, between the axis of rotation of the target and the normal to the crystal planes.

A third effect is that of gas adsorbed on the target. This was studied by strongly heating the target, allowing it to cool again, and then repeating the observations. Most of the spurs were ultimately made much more intense by several repetitions of this process; others would disappear and reappear, or disappear entirely, and were regarded as due to scattering by the adsorbed gas.

The most suggestive aspect of the phenomenon was the similarity, noted in the first quotation, to the Laue beams produced when x rays strike a crystal. As Davisson and Germer put it in

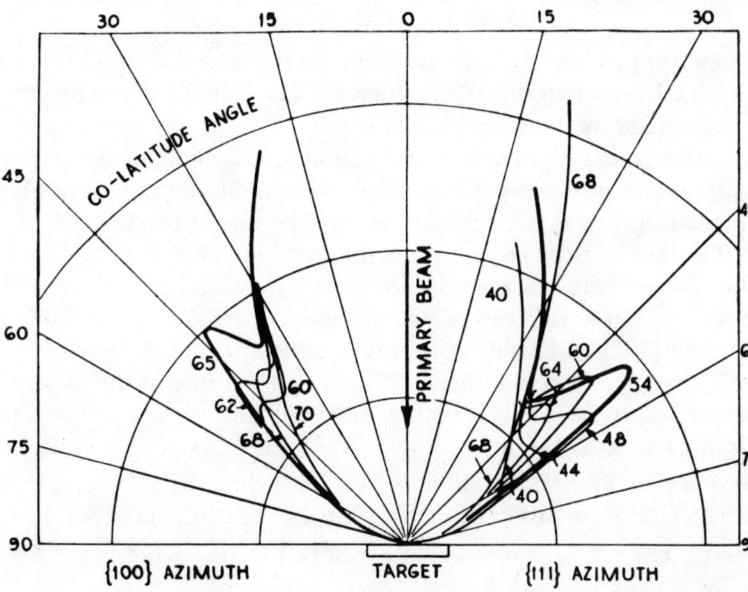

FIG. 10.7 Scattering curves showing the occurrence of two pronounced maxima. Right side: {111} azimuth; left side: {100} azimuth. Each curve is labeled with the bombarding potential in volts. [*Phys. Rev. 30*, p. 716 (1927), Fig. 10.]

their note in *Nature,* "If the incident electron beam were replaced by a beam of monochromatic X-rays of adjustable wave-length, very similar phenomena would, of course, be observed. At particular values of wave-length, sets of three or of six diffraction beams would emerge from the incident side of the target." They then proceeded to associate x-ray wavelengths with ten of the thirteen beams they had found at that time. They go on:

These results are highly suggestive, of course, of the ideas underlying the theory of wave mechanics, and we naturally inquire if the wave-length of the X-ray beam which we thus associate with a beam of electrons is in fact the h/mv of L. de Broglie. The comparison may be made, as it happens, without assuming a particular correspondence between X-ray and electron beams. . . . The wave-lengths of all possible X-ray beams satisfy the optical grating formula $n\lambda = d\sin\theta$, where d is the distance between lines or rows of atoms in the crystal—these lines being normal to the azimuth plane of the beam being considered.

They applied the grating formula to the electron beams, and obtained wavelengths that generally correlated well with the values given by de Broglie's relationship.

There is no reason to repeat the analysis for all 30 of the sets of spurs found in the experiment, nor even to tabulate the resulting correlation; but it is instructive to consider one or two cases.

The first step is to compute the grating spacing d, which is the spacing between rows of atoms in the surface. As can be seen from Fig. 10.5, this distance is different for different azimuths: For the {111} and {100} azimuths it is the altitude of the elementary triangle, while for the {110} it is half the side of the elementary triangle. We will consider only the former case. Since the triangle is equilateral, its altitude d is $\frac{1}{2}\sqrt{3}$ times the length s of one side. Figure 10.4 shows that s, in turn, is half the diagonal of the face of the unit cube; if the edge of the cube is of length a, then $s = a/\sqrt{2}$ and $d = a\sqrt{3}/2\sqrt{2}$. Substitution of the value 3.51 Å for a gives $d = 2.15$ Å.

Let us now use this value in treating the two beams indicated in Fig. 10.7. The one shown on the right-hand side of the figure has its maximum development at 50° and 54 volts. If we assume that it is a first-order beam, then $n = 1$ and the grating formula gives a wavelength $\lambda = d\sin\theta = 2.15$ Å $\times$ sin50° $= 2.15$ Å $\times 0.766 = 1.65$ Å. This is to be compared with the value obtained from de Broglie's formula $\lambda = h/mv$, with m the mass of the electron and v its velocity. The velocity is determined by the accelerating potential V, so that $\frac{1}{2}mv^2 = Ve$, where e is the charge of the electron. Multiplying this equation by $2m$ gives $m^2v^2 = (mv)^2 = 2m\,Ve$, or $mv = (2m\,Ve)^{1/2}$, and de Broglie's formula becomes $\lambda = h/(2m\,Ve)^{1/2}$. Planck's constant is $h = 6.62 \times 10^{-27}$ erg second, the mass of the electron is $m = 9.11 \times 10^{-28}$ gram, and the charge of the electron is $e = 4.80 \times 10^{-10}$ electrostatic unit; for use with these units, the accelerating voltage must be converted to electrostatic units by use of the relationship 1 volt = 1/300 electrostatic unit, and the wavelength is then given in centimeters. Thus $\lambda = [6.62 \times 10^{-27}/(2 \times 9.11 \times 10^{-28} \times 54 \times 4.80 \times 10^{-10} \times 1/300)^{1/2}]$ cm $= 1.67 \times 10^{-8}$ cm $= 1.67$ Å. The agreement is quite satisfactory. Similarly, for the 65-volt beam reaching a maximum at 44°, the grating formula gives $\lambda = 1.49$ Å, while de Broglie's formula gives $\lambda = 1.52$ Å, again in reasonable agreement.

A complete analysis, of course, is much more involved. If the grating equation were the only factor to consider, there would be no explanation of the fact that the beams do not simply shift as the voltage is varied, but grow and disappear. The fact of the matter is that a crystal is not just a two-dimensional grating but a three-dimensional array, and account must be taken of possible interference between waves reflected from successive layers. This was done in the article. The authors were still unable to refrain from associating a Laue beam with each electron spur, and making their comparison on this basis, so that there are quantitative flaws; but the fundamental concept was clearly valid.

At almost the same time, Sir George Thomson, son of Sir J. Thomson of the Thomson atom, was performing a related experiment in England, using transmission of high-energy electrons (several kilovolts) through metal foils rather than reflection of low-energy electrons from single crystals.[7] The analysis of his experiment is considerably more complicated than that of the Davisson-Germer experiment, but the result was an independent confirmation of de Broglie's proposed relationship between wavelength and momentum for electrons.

There was still one question unanswered, however. It was possible to believe that the wavelike behavior revealed in these experiments was peculiar to electrons. Would other matter display the same properties? The answer came in 1930, when Otto Stern and his co-workers established that atoms and molecules also had wavelike attributes, with the wavelength given by de Broglie's relationship.

In contrast to Davisson and Germer, Stern and his colleagues had deliberately set out to test de Broglie's hypothesis. In a paper published in 1926, in which he introduced and gave the background for a series of "Investigations in molecular beam methods," he said,

> A question of the greatest fundamental significance is that of the real existence of De Broglie waves, that is, the question of whether for molecular rays, in analogy with light rays, scattering and interference effects are to be observed. Unfortunately, the wavelengths calculated according to De Broglie even under the most favorable conditions . . . barely come up to 1 Å. Nevertheless, the possibility for experiment to demonstrate such effects for molecular beams appears not to be excluded.

The "most essential parts" of a molecular beam apparatus, as described in the second paper of the series, written in collaboration with F. Knauer, "are the oven and the oven opening, the image aperture and the detector screen." For the experiments to be discussed here, the oven is replaced by a source of gas and the screen by another type of detector, but the essential concepts are the same.

> From the oven opening, the molecules enter the vacuum in a wide cone; there, if no collision takes place, they follow straight-line paths (apart from the influence of gravity, which in our investigations is of no importance . . .). The image aperture cuts off from the broad cone a narrow, almost parallel beam. At the detector surface the molecules of the ray are condensed. . . .
>
> According to an unobjectionable ray picture the molecular beams propagate exactly like light rays in the casting of a shadow.

They also could be reflected, like light rays, from a smooth surface.

Because the mass of the particle appears in the denominator of the de Broglie expression for wavelength, it was desirable to use the lightest possible molecules, hydrogen and helium, to test it. These could not be detected by the condensation method. It was not until 1929 that a suitable detection method was developed and described, again by Knauer and Stern.

> The principle of the method is that the beam of molecules strikes a slit in an otherwise closed chamber. As a result, the pressure in the chamber increases up to the point where just as much gas flows out of the slit [into the surrounding vacuum] as the molecular beam carries in.

For a given chamber and slit, the outward flow will be proportional to the pressure in the chamber. The steady-state pressure, therefore, is a direct measure of the incident intensity. Knauer and Stern measured this pressure by "a hot-wire manometer after Pirani." [8] They succeeded in improving the sensitivity of this type of gauge to the point that they could measure pressures of the order of 10^{-5} millimeter of mercury to a precision of a tenth of a percent.

Even then, the results came as a sort of series of successive approximations. The first efforts, described by Knauer and Stern in 1929, attempted to make use of the fact, well known in optical spectroscopy and already used with x rays, that if light is incident on a grating at an angle θ from the normal, the effective spacing is less than the actual by a factor $\cos\theta$. By use of grazing incidence, which means θ near 90°, the effective spacing of an ordinary mechanically ruled grating can be made of the order of an angstrom. Such an effective spacing should have produced detectable interference maxima analogous to fringes in the optical case. The attempt, however, was unsuccessful: "Although several times we believed that we had found indications of a maximum, we did not succeed in establishing its certain existence. The difficulty lies in the fact that if the scattering maximum is located too close to the reflected ray, the very small but definite lack of sharpness in the latter makes the measurement uncertain, whereas if the scattering maximum is located too far from the reflected ray, its intensity becomes too small." Nevertheless, the results were encouraging: Not only was there the suggestion of a diffraction maximum, as noted in the foregoing quotation, but the reflection coefficient (reflection gratings were used) varied with angle of incidence and with gas temperature (and hence speed and wavelength) in the way that would be expected from a wavelike behavior.

Accordingly, studies were undertaken using a crystal lattice as a grating. "Since the lattice constant in this case is in order of magnitude a few times 10^{-8} centimeter, we could work at larger angles of incidence[9] (5 to 45°), which from the experimental viewpoint meant an improvement." A diagram of the apparatus is shown in Fig. 10.8. *O* is the "oven" slit, to which gas was led through the tube *a*; D_1 could be either a Dewar vessel, for cooling the gas to liquid-air temperature, or a heater. The image slit is denoted by *Aa* in the plan view, *Ab* in the side view. K_2 is the crystal, *Af* the detector slit, and R_1 the tube connecting the detector chamber to the measuring manometer. The positions of the detector and the crystal could be set by means of the brass slip joints S_1 and S_2, respectively, which were accurately coaxial with their common axis lying in the surface of the crystal.

The results were rather disappointing; in particular, the hoped-for diffraction patterns still could not be definitely established.

> Nevertheless, in all measurements there appeared the remarkable result that for small angles the reflection was better the higher the temperature of the ray, and therefore the smaller the wavelength of the corresponding de Broglie wave. For larger angles of incidence (30°) the reverse is true. . . .
>
> Apart from the fact that such a sharp reflection is not to be understood on the grounds of the classical theory and can only be interpreted by recourse to the wave nature of the beam of molecules, also the above-described behavior of the reflection coefficient is only to be understood on the grounds of the wave theory.

But Stern was not to be deterred by the lack of complete success. Even though "Mr. Knauer, to my great regret, for the sake of his own studies wished to discontinue the collaboration in the research," he carried on, first alone and then in collabora-

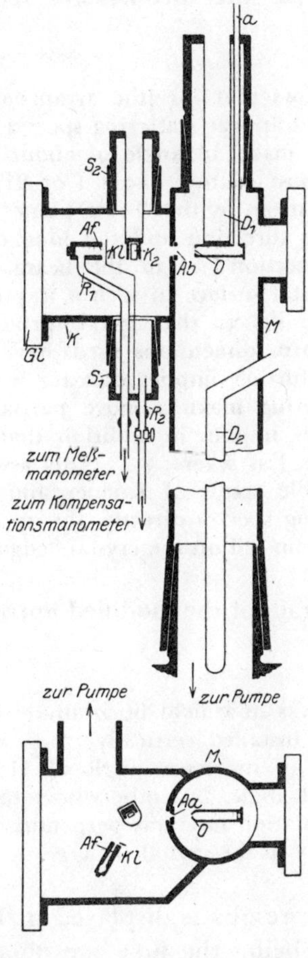

FIG. 10.8 First version of the apparatus used by Knauer and Stern for studying the scattering of molecular beams by a crystal. [*Z. Physik 53, p. 783 (1929)*, Fig. 2.]

tion with I. Estermann, together with whom he published conclusive results in 1930. Apparently one of the most productive things he had done in the meantime was to study much more thoroughly the theory of diffraction by a two-dimensional grating such as is realized by a crystal surface (as contrasted with the one-dimensional character of a ruled grating), for the first major topic discussed in the 1930 paper is how the properties of lattice diffraction account for the inconclusive nature of the earlier results.

> The theory . . . shows that for [the arrangement used by Knauer and Stern] there should appear scattering spectra that for the expected de Broglie wavelength make an angle of about 8 to 9° with the reflected ray and lie almost (within about 1 or 2°) in the beam plane. In this connection we mean by the "beam plane" the plane that is determined by the beam direction and the long edge of the rectangle formed by the cross section . . . of the beam. The beam in these studies fell on the crystal surface in such a way that the long edge of this rectangle was parallel to the crystal surface. . . . Such maxima must, with the apparatus dimensions used, have also gotten into the detector. The theory further implies that for a small rotation of the crystal the said scattering maxima move partially out of the beam plane. . . . The theory implies in addition that for larger angles of incidence this effect is less severe. . . . This assumption immediately explains the remarkable result of Knauer and Stern. . . . In order to study these scattering spectra directly, the apparatus was rebuilt in such a way that the beam fell on the crystal "edgewise."

Figure 10.9 is a diagram of the modified apparatus. The essential differences were

> . . . that the crystal was now held horizontally in the crystal holder, whereas earlier it was installed vertically . . . , and the beam was accordingly inclined at an incidence angle of 11½°. . . . The detector was tilted at an equal angle, but otherwise retained its arrangement such that its axis of rotation now was perpendicular to the crystal surface, while before it lay in the crystal surface.

The nature of the results is displayed in Fig. 10.10, with the curve labeled "III" being the first one obtained with the new arrangement. "They definitely show the expected scattering maxima." The absence of maxima in Curve IV was also in accord with theory. However, "The shift of the scattering maxima with temperature to be expected according to the de Broglie formula[10] . . . is in the proper direction, but substantially too small."

This flaw, in turn, was traced to inadequate adjustability of the apparatus. For one thing, the structure supporting the "oven" was changing its dimensions with temperature. More important, it was found that sometimes the crystal surface was not truly perpendicular to the axis, so that the true directions of incidence and scattering were not what were measured. A further modification of the crystal holder was made to permit adjustment of the latter aspect. It was such that in addition, the angle of

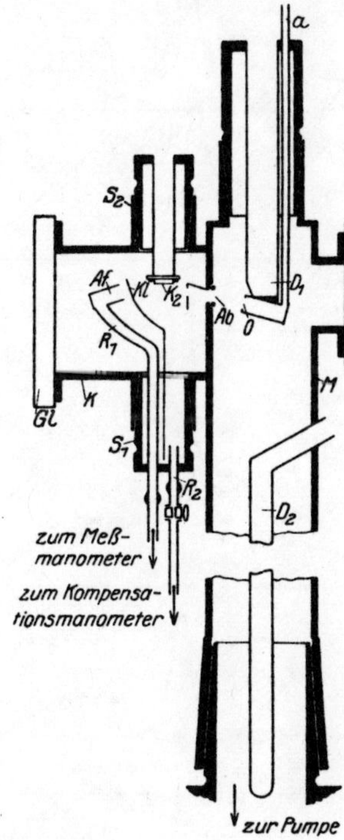

FIG. 10.9 Later version of the apparatus for studying the scattering of molecular beams. [Z. *Physik 61*, p. 98 (1930), Fig. 7.]

incidence could readily be varied through several degrees; this permitted a considerably more extensive search to be made. In fact, "For each position of the detector we looked for the 'most favorable tilt,' that is, the tilt for which the intensity of the scattered beam reached its maximum. . . ."

Estermann and Stern carried the work with rocksalt just enough further to convince themselves that the improved apparatus was permitting genuine and accurate measurements of diffraction from a two-dimensional grating. "Meanwhile," they

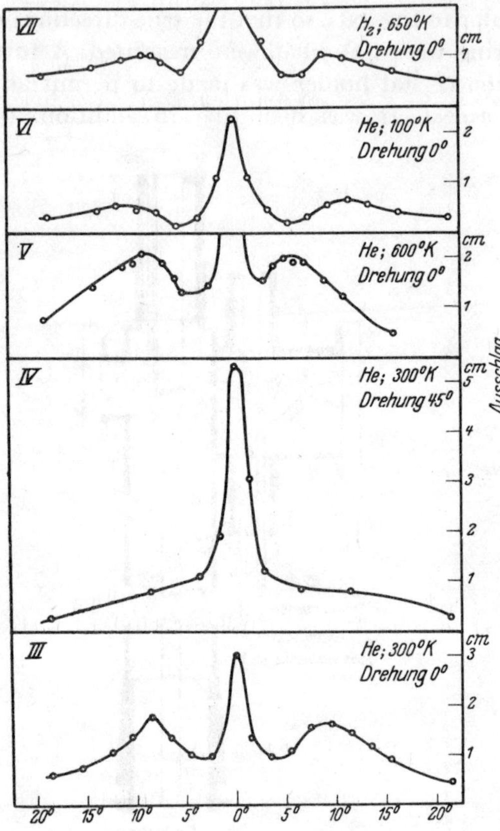

FIG. 10.10 Scattering of He and H_2 on rocksalt. All curves except IV are taken with the crystal so oriented that the plane of incidence, or a plane perpendicular to it, contains only atoms of a single species; curve IV is taken with the crystal rotated in its own plane through 45° (*Drehung* means "rotation"), so that the atoms in the plane of incidence are sodium and chlorine alternately. [*Z. Physik 61*, p. 99 (1930), Fig. 8.]

write, "we had found that studies with the LiF lattice result in a much simpler and clearer picture," since "We obtained substantially better reflection, sharper beams and more intense maxima. . . ." The remainder of the work was accordingly done with lithium fluoride.

Typical results are shown in Fig. 10.11, and several sets of results are summarized in Table 10-1. The agreement is generally good; particularly interesting is the fact that the maximum falls at the same position for helium at 290° as for hydrogen (half the mass of helium) at 580° (twice 290°).[11] There was still, however, a systematic deviation for large angles, corresponding to long wavelengths. Part of this was ascribable to the nature of the experimental arrangement, whereby "for large angles, the

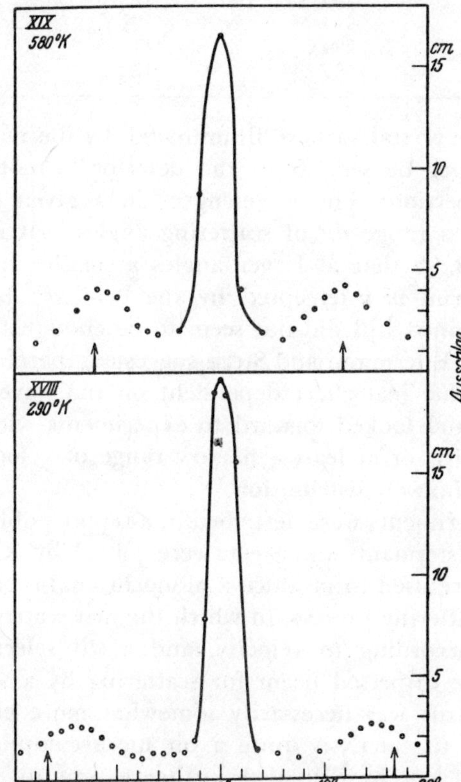

FIG. 10.11 Scattering of H_2 on lithium fluoride. The arrows show the positions of the maxima calculated on the basis that all the molecules have a single velocity, that at which the Maxwell distribution for the appropriate temperature has its maximum. [Z. *Physik* 61, p. 107 (1930), Fig. 16.]

TABLE 10-1. *Comparison of Observed and Calculated Positions of Maxima for H_2 and He Scattered from LiF*. [Adapted from I. Estermann and O. Stern, *Z. Physik* **61**, 95 (1930), Table 4.]

Gas	Temperature (°K)	Location of maximum (deg) Calculated	Observed
He	100	21	15½
	180	15½	14½
	290	12	11½
	590	8¾	9
H_2	290	17	17
	580	12	11

whole of the crystal surface 'illuminated' by the molecular beam can no longer be seen from the detector." Another part was due to dispersion: The wavelengths in a given range $d\lambda$ are spread over a range $d\beta$ of scattering angles that increases with increasing β, so that at larger angles a smaller fraction of the scattered beam is intercepted by the detector. But these two effects combined still did not seem to be enough to explain the discrepancy. Estermann and Stern suggested, therefore, that there might be some real effect dependent on the wavelength of the molecules, and looked forward to experiments with beams of a single velocity, or at least a narrow range of velocities in place of the full Maxwell distribution.

Such experiments were described in a report published in 1931, for which Estermann and Stern were joined by R. Frisch. Two methods were used to produce a monochromatic beam. One was a double-scattering process, in which the first scattering dispersed the beam according to velocity, and a slit selected a narrow range of the dispersed beam for scattering by a second crystal. The apparatus was necessarily somewhat more elaborate than before, and the analysis quite a bit more complicated; so this method will not be discussed further except to state that the results were completely as expected.

The second method was much more direct, a mechanical "monochromatization." (A footnote in the original report comments that "Mr. Estermann on account of a trip to America was no longer able to participate in the final measurements.")

For the mechanical monochromatization the helium beam was sent through a system of two toothed wheels fixed on the same axle, rotating rapidly. The toothed wheels had a diameter of 19 cm and were set 3.1 cm apart. Each of them had in the circumference 408 equidistant radial saw cuts of breadth 0.4 mm and depth 5½ mm. The toothed wheels were not displaced relative to one another, so that exactly two corresponding slits simultaneously crossed the beam path (lying parallel to the axis of rotation). Thus if the wheels turned very slowly, molecules of all velocities could pass through and the beam was merely weakened in the ratio (slot width):(slot separation). For somewhat faster rotation of the toothed wheels, the slower molecules that had gone through one gap in the first wheel could no longer reach the corresponding gap in the second wheel. For still faster rotation, this was also the case for the speedier molecules, but now the slower molecules could get through the next following slit in the second toothed wheel. The faster the wheels rotated, the greater also was the speed of the molecules that were selected in this way. If we denote by ν the number of turns [per unit time], and by z the number of slits, then $1/vz$ is the time in which the toothed wheel advanced by one slit. If we neglect the width of the slit, then only molecules with velocity v can get through the next following slit that take exactly that time for covering the separation distance l between the two wheels. Thus v is determined by the equation $l/v = 1/\nu z$, $v = lz\nu$. In our case we had $l = 3.1$ cm (separation between facing surfaces 3.0 cm, plus 0.1 cm thickness of the wheel), $z = 408$, therefore $v = 1265\nu$ cm/sec.

The resulting monochromatic beam was analyzed by scattering from a crystal of lithium fluoride, much as in the previous work.

A diagram of the apparatus is shown in Fig. 10.12. The nozzle of the oven is at *Or*; the beam is defined by a channel-shaped slit s_1. The toothed wheels are seen edgewise at Z_1 and Z_2, with the axle *A* carried in ball bearings and fastened by a flexible coupling to a second axle which passed through an oil seal in the vacuum housing and carried the drive pulley. The detectors *Af* and *Afk* were made by milling rectangular slots in a brass plate that had been ground plane, and then covering the slots with glass plates. The bearing for the axis by which the crystal was rotated was also milled into the plate, ensuring coplanarity of the axis and the detectors. A bit of gadgetry enabled the crystal to be displaced out of the beam, which then fell directly on the compensation detector *Afk*.

The first step was to test the action of the velocity selector

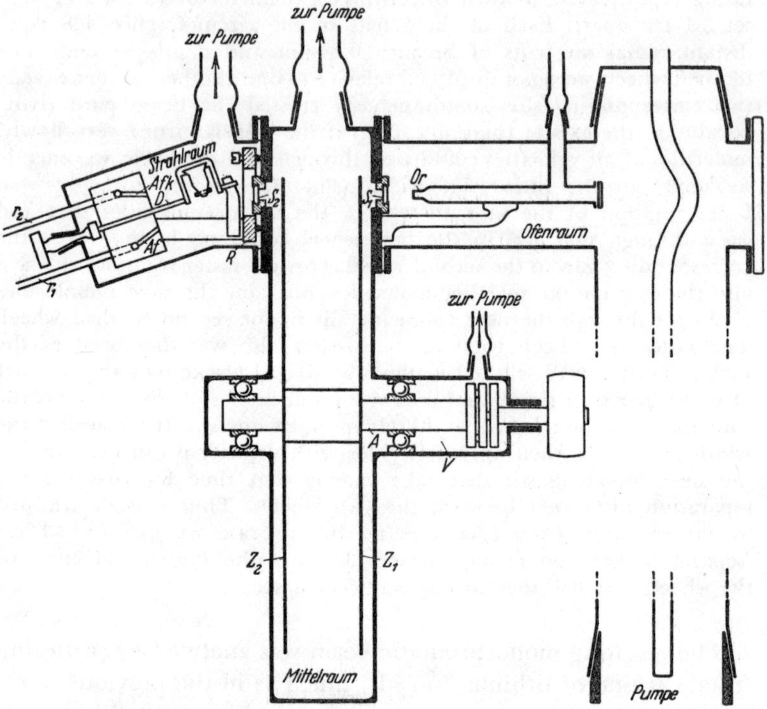

FIG. 10.12 Schematic diagram of the "mechanical monochromator" (velocity selector) and associated scattering crystal and spectrometer of Estermann, Frisch, and Stern. [Z. *Physik* 73, p. 358 (1931), Fig. 11.]

by measuring the intensity of the direct beam as a function of speed of rotation of the wheels. According to the kinetic theory of gases, the result should have been as shown in Fig. 10.13(a). Instead, it showed an initial rise and then a decrease, as in Fig. 10.13(b). This behavior was due to a misadjustment of the slits s_1 and s_2 such that the beam was not traveling exactly parallel to the axle. Part of this could be corrected, but it was not possible to achieve exact adjustment. "We contented ourselves with this degree of accuracy of adjustment and compensated for this error by calculating the velocity v of the molecules from the formula

$v = lz\nu \cdot 0.969 = 1226\nu$ cm/sec," the factor 0.969 being calculated from the degree of residual misadjustment and the dimensions of the wheels.

All was now ready for the actual measurements, the determination of the scattered intensity as a function of angle for various values of rotational speed. The raw curves obtained are shown in Fig. 10.14. The top curve is for 3 revolutions per second, slow enough that practically all molecules could pass through the

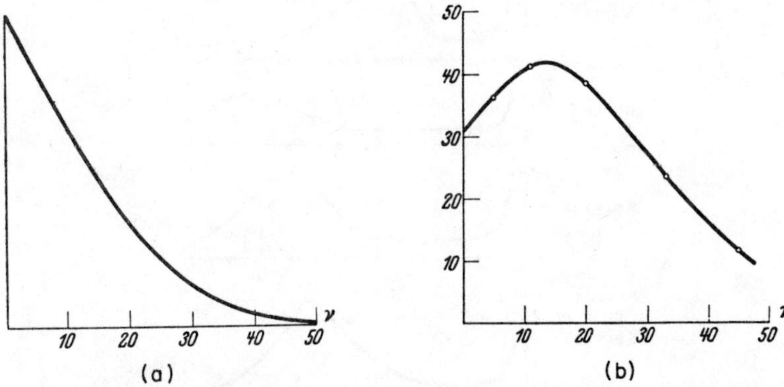

FIG. 10.13 (a) Ideal curve of intensity of the direct beam through the velocity selector as a function of rotational speed ν. (b) Actually measured intensity of the direct beam as a function of selector speed. [Z. *Physik* 73, p. 359 (1931), Figs. 12 and 13.]

slits, and shows just the distribution of speeds expected from kinetic theory, the Maxwell distribution. In the next curve, some of the slower molecules are blocked off, and the distribution is correspondingly shifted to higher speeds, smaller de Broglie wavelengths, and smaller scattering angles. The remaining curves

> . . . show the scattering curves taken with higher rotational speeds, thus with monochromatization. As can be seen, the scattering maximum approaches steadily closer to the reflected beam for increasing rotational speed, as the de Broglie wavelength becomes shorter, the faster atoms one takes.
>
> From the rotational speed ν, the velocity was calculated according to the formula $v = 1226\nu$ cm/sec, and from that, according to the de

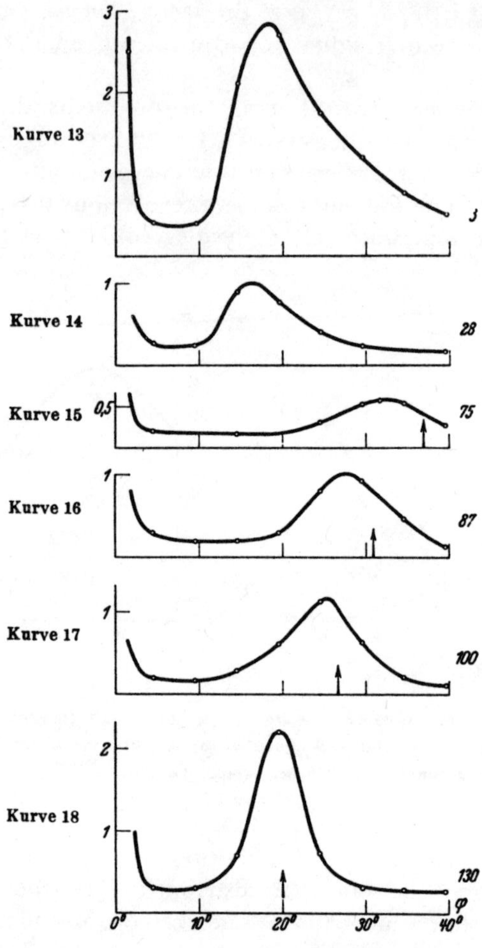

FIG. 10.14 Raw curves of scattered intensity versus scattering angle for several values of rotational speed of the velocity selector, given in revolutions per second by the numbers on the right. [Z. *Physik 73*, p. 362 (1931), Fig. 15.]

Broglie relationship, the corresponding wavelength $\lambda = h/mv = (80.5/\nu) \times 10^{-8}$ cm. The scattering angle calculated from this wavelength . . . is designated by an arrow. The measured maxima all lie at somewhat too short waves. That is to be expected, since for the rotational speeds used we find ourselves on the increasing side of the Maxwell curve, that is, in the velocity interval selected the fast atoms (short wavelengths) are heavily overabundant.

This is shown more explicitly in Fig. 10.15. The solid curve shows the overall Maxwell distribution, from which the velocity selector passes only the segment defined by the upper and lower velocity limits represented by the vertical lines. From the

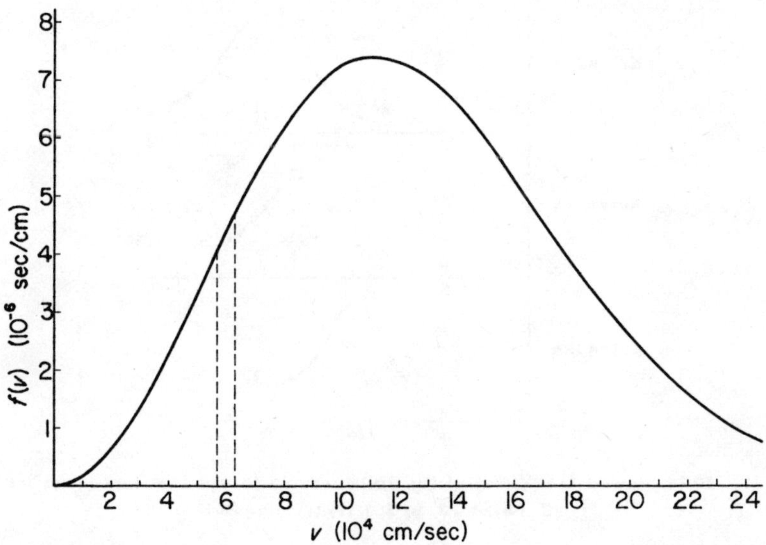

FIG. 10.15 Nature of the "cut" taken from the Maxwell distribution by a mechanical velocity selector. The solid curve represents the distribution function $f(v)$, with the property that $f(v)dv$ gives the fraction of the molecules in the beam having speeds between v and $v + dv$. The vertical (dashed) lines indicate the manner in which the velocity selector "cuts" the distribution.

meaning of the distribution curve, it follows that such a segment contains many more molecules with speeds near the upper limit than near the lower.

We have accordingly corrected [the lowest four curves of Fig. 10.14] by the method that we divided each ordinate value by the ordinate value corresponding to the same abscissa of the nonmonochromatized [top curve of Fig. 10.14], thus so to speak reducing to equal incident intensity of all wavelengths. . . . For the curves so obtained [Fig. 10.16], the maxima lie within the accuracy of the measurement at the calculated positions. . . .

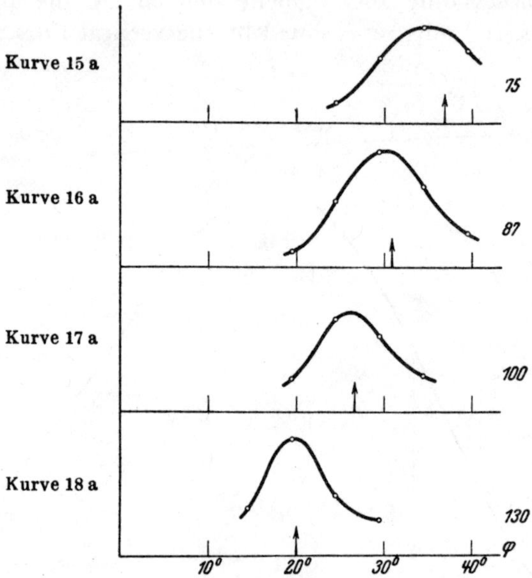

FIG. 10.16 Lowest four curves of Fig. 10.14 reduced to equal incident intensities. [Z. *Physik* 73, p. 363 (1931), Fig. 16.]

The shapes of the curves show that the positions of the maxima can be determined fairly accurately for high speeds of rotation. These speeds, moreover, bring the selected interval near the maximum of the Maxwell distribution. "Since in this region also the correction just discussed is small, a measurement was carried out at 133.3 revolutions per second with special care, in order to test the de Broglie relationship $\lambda = h/mv$ also numerically as exactly as possible." Measurements were made on two separate days with different crystals; in each case, the reflected ray and the maxima on both sides of it were measured. The maxima were found to be symmetrically located about the reflected ray to within 0.1°; the raw maximum angle was 18.9°

for one run, 19.0° for the other, for an average of 18.95°. The reduction to equal incident intensity gave a shift of 0.5°, to 18.45°, corresponding to a wavelength of 0.600×10^{-8} cm. The rotational speed of 133.3 revolutions per second gave a velocity of 1.635×10^5 cm/sec, corresponding to a wavelength of 0.604×10^{-8} cm. The discrepancy of two-thirds of one percent was well within the experimental uncertainty of one to two percent, and the de Broglie hypothesis was quantitatively verified. The successful operation of the experiment, of course, had verified the qualitative aspects.

Apparently, then, matter has wavelike properties which can be made observable by proper choice of parameters. For a time, not surprisingly, this and the analogous behavior of light were regarded as paradoxical. Gradually, however, acceptance has grown for the following interpretation: Neither the classical concept of "particle" nor that of "wave" can be correctly extrapolated to the realm of the very small. Rather, there is a third entity for which we have no single name, which acts in some ways like a classical particle and in some ways like a classical wave. It is this "stuff" of which the universe is made.

FOOTNOTES

[1] These relationships would lead to a wave velocity greater than that of light, having no clear connection with the speed of the piece of matter. The solution to this difficulty is to assume that the matter corresponds not to a single wave but to a group of waves. The individual waves travel with a velocity greater than that of light; but the group, and with it the energy and momentum, travels at the speed of the matter.

[2] This is roughly the speed of the tip of the minute hand on a clock about four to five inches in diameter.

[3] A preliminary account was published in the British journal *Nature* earlier in the year.

[4] A person to whom this happens is said to show *serendipity*. The word is derived from a story by Horace Walpole, "The Three Princes of Serendip," in which the princes of the title had this trait.

[5] The scattering was actually backward, and present usage favors giving the supplements of these angles. We are following the terminology of the original article.

[6] It proved impracticable to measure and hold constant the bombarding current itself. What was kept constant was the current to one of the electrodes of the electron gun; but this still permitted the bombarding current to vary somewhat as the overall accelerating voltage was changed.

[7] Sir George has given some interesting recollections about his work: G. P. Thomson, *American Journal of Physics* **29,** 821 (1961). See also G. P. Thomson, *Physics Today* **20,** No. 5, 55 (1967), where his work and that of Davisson and Germer are discussed in the context of an overall history of the concept of the electron.

[8] The Pirani hot-wire manometer is based on the fact that the equilibrium temperature of an electrically heated wire depends in large part on the rate at which heat is carried away by the surrounding gas, and hence on the density and therefore the pressure of the latter.

[9] Note that here, as in x-ray research, the angle of incidence is conventionally measured from the surface, rather than from the normal as it is in optics.

[10] In the de Broglie relation $\lambda = h/mv$, the denominator can be written as $(m^2v^2)^{\frac{1}{2}} = (2mE)^{\frac{1}{2}}$; for a Maxwell distribution, the most probable energy is proportional to kT, so that $\lambda_{max} = h/(CmT)^{\frac{1}{2}}$ where C is a constant. Thus an increase in temperature (of the oven) should produce a decrease in wavelength.

[11] See the formula derived in footnote 10, which shows that simultaneously doubling the temperature and halving the mass should leave the wavelength unchanged.

BIBLIOGRAPHY

The Davisson-Germer experiment is reported in C. J. Davisson and L. H. Germer, *The Physical Review* **30,** 705–740 (1927). See also *The World of the Atom,* Vol. 2, pp. 1137–1142 and 1144–1165.

Thomson's work is presented in G. P. Thomson, *Proceedings of the Royal Society of London, Series A* **117,** 600–609. See also *The World of the Atom,* Vol. 2, pp. 1137–1144.

The essential papers from Stern's group are F. Knauer and O. Stern, *Zeitschrift für Physik* **53,** 766 and 779 (1929); I. Estermann and O. Stern, *ibid.* **61,** 95–125 (1930); I. Estermann, R. Frisch, and O. Stern, *ibid.* **73,** 348–365 (1931). These are papers Nos. 10, 11, 15, and 18 of the series mentioned on p. 118; all are in German.

Appendix A

An electrometer is an instrument for measuring potential differences without drawing significant currents. The quadrant electrometer consists of a metal "pillbox" divided into four quadrants by means of radial cuts, as in Fig. A.1. A light metal vane, designated by V in the figure, is suspended in a horizontal plane within the pillbox by means of a torsion fiber, which also carries a light mirror. The vane is free to rotate about a vertical axis, and its angular position can be measured by a beam of light reflected from the mirror.

The quadrants are electrically connected in pairs, pair A con-

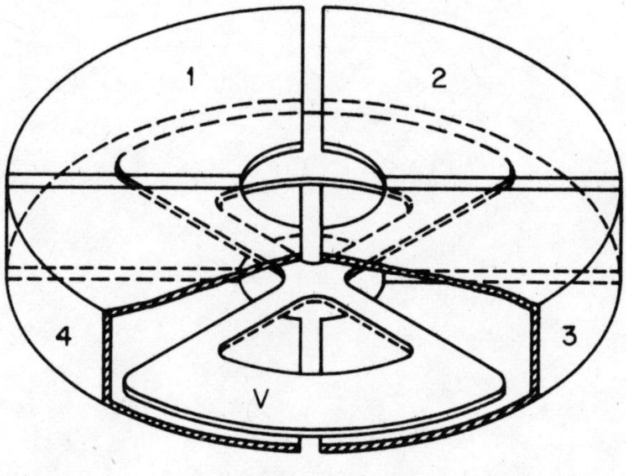

FIG. A.1 Diagram of the working parts of a quadrant electrometer.

sisting of quadrants 1 and 3 and pair *B* of quadrants 2 and 4. To use the instrument, the vane is charged to some chosen potential, and the potential difference to be measured is applied between the pairs of quadrants; the vane then takes up an angular position that is determined by the potentials involved and the stiffness of the suspension fiber.

The instrument can also be used to measure very small currents by making use of the fact that the charging or discharging of a capacitance changes the potential difference across it. Such use was made by early workers in many fields. The capacitance is essentially just that of the quadrants themselves. The pairs of quadrants are initially connected together, one pair being permanently connected to ground ("earth," in British usage), and the other to the current-collecting electrode. The connection between the quadrant pairs is then opened, and the current charges the capacitance, changing the potential difference between the quadrant pairs and thus causing the vane to turn at a rate which is (approximately) proportional to the current.

The quadrant electrometer is now rarely used, as electronic devices have been developed that do as well or better with less trouble.

Appendix B

NUMERICAL NOTATION FOR CRYSTALS

To understand the numerical notation used in specifying planes and directions in a crystal,[1] imagine a set of coordinates with their origin at one corner of the basic cell and the axes along the cell edges. (For some types of crystals, the axes will not be mutually perpendicular. Also, for some types of crystals, the choice of which axis lies along which cell edge is significant and a particular one is standard. For a cubic crystal, neither of these features appears.) Any plane of atoms in the crystal intersects each axis a whole number of cell edge lengths from the origin, and any particular one of the planes would be completely specified by the three numbers of cell edge lengths giving the intercepts. However, all planes parallel to each other are equivalent, and it is preferable to use a set of numbers which will be the same for all planes of a parallel family. Such a set is obtained by taking the reciprocals of the set just described, and multiplying them by the smallest factor that will give three integers. For example, a plane that intersects the x axis two units from the origin, the y axis three units, and the z axis one unit would give the set of reciprocals $\frac{1}{2}$, $\frac{1}{3}$, 1; multiplication by six gives the indices {326}. If one index is zero, the planes of the family are parallel to the corresponding axis; thus, the {110} planes are parallel to the z axis, and the {010} planes are parallel to the xz plane. Finally, a line is specified by the indices of the family of planes to which it is perpendicular.

FOOTNOTE

[1] Further details and additional references are given by Elizabeth A. Wood, *Crystals and Light,* MOMENTUM Book No. 5 (Van Nostrand, 1964), Chap. 3.

Index

Entries followed by "n" refer to footnotes; by "b," to bibliography.

Angular momentum, atomic, 88 ff., 96n.
Atomic concept, atomic hypothesis (*see also* Molecular hypothesis), 3, 23, 37, 38, 55
  opposition to, 37, 54n.
Avogadro's number, 19, 37, 40, 42, 50, 51

Becquerel, Henri, 23
Black body, defined, 4
Blackbody radiation (*see* Cavity radiation)
Bohr, Niels, 75, 88, 96n.
Boltzmann, Ludwig, 8, 19
Brownian motion, 37 ff., 50

Cavity radiation
  absolute character, 6
  equivalence to blackbody radiation, 6
  Rayleigh's law, 11, 19
  temperature dependence, classical, 9
  Wien's spectral function, 9, 14, 19
Classical physics
  continuum nature, 3, 37
  dynamics, 1
  electromagnetism, 1, 78, 97
  thermodynamics, 1
Collisions, elastic and inelastic, 71 ff.
Collisions, ionizing, 70
Compton, Arthur H., 8, 83, 98 ff.
Contact emf, contact potential difference, 79, 82, 87n.
Crystal lattice, 107, 120, 137, 138
  of nickel, 110 ff.
Curie, Marie and Pierre, 24

Davisson, C. J., 105 ff.
de Broglie, Louis, 105, 116
de Broglie wavelength, 105, 107, 117, 120, 122, 129, 130, 132, 134n.
de Broglie waves, 118, 120 ff.
Density of emulsion grains, 44
Diffraction of electromagnetic waves, 97, 100, 107
Dispersion, 126
Dynamid, 56

Einstein, Albert, 54n.
Brownian motion, 50
  photoelectric effect, 76 ff., 86, 87
  relativity theory, 4, 20n.
Electron
  charge, 19, 84
  constituent of atoms, 3n., 23, 55, 56, 69, 72
  in metals (photoeffect), 77
  scattering of electromagnetic radiation, 97 ff.
Emanation, 25 ff., 62, 66

Energy values of atom, discrete character, 75
Estermann, I., 121 ff.
Extrapolation, 2
  failures, 2, 75, 76, 89, 97

Fractional centrifuging, 43
Franck, James, 69 ff., 96n.
Frisch, R., 126 ff.

Gas law, 40
Geiger, Hans, 57 ff.
Gerlach, Walther, 94 ff.
Germer, L. H., 105 ff.
Grating diffraction, 100, 107, 116, 119 ff.
Grating, multidimensional, 117, 121 ff.

Hertz, Gustav, 69 ff.
Hughes, Arthur L., 80

Ionization (*see also under* Radioactivity)
Ionization energy, 72
Ionization potential (*see also* Ionization energy), 69 ff.
Interference of waves, 103, 107, 117, 118

Kelvin, Lord, 3n., 55
Kinetic theory (*see also* Molecular theory), 3, 105, 128
Kirchhoff, G. R., 6, 21b
Knauer, F., 118 ff.

Laue beams in x-ray scattering, 107, 115, 117
Laue, Max von, 97
Lenard, Philipp, 55
Lummer, O., 7 ff.

"Machine shop in vacuo," 80
Magnetic moments of atoms, 89
Marsden, Ernest, 57 ff.
Maxwell distribution of molecular speeds, 126, 129 ff.
Maxwell, James C., 21n., 56
Maxwell's theory (electromagnetism), 97
Michelson, A. A., 3n., 4
Millikan, Robert A., 54n., 77 ff., 84
Molecular beams, 118 ff.
Molecular hypothesis, 53
Molecular theory, 40, 42

Nagaoka, H., 56, 67n., 67b

Optical spectra, 55, 69
Osmotic pressure, 40, 41

Paschen, F., 8
Perrin, Jean, 37 ff.
Photoelectric effect, 74
  critical frequency, 76, 79, 84 ff.
Photon (*see* Quantum of radiation)
Planck's constant (*see also* "Quantum of action"), 19, 72, 77, 84, 87, 88, 99, 107, 117
Precession, of atomic angular momentum, 90
Pringsheim, E., 7 ff.

"Quantum of action," 4
Quantum of radiation, 76, 77, 78, 87, 98, 103n.
Quantization
  angular momentum, 88, 96n.
  energy, 19, 75, 77, 87, 88

Radiation, excitation by collisions, 72 ff.
Radioactivity
  absorption measurements, 25
  decay, 27, 28, 33, 35n., 64, 66
  "excited," 27, 28, 33, 34
  ionization measurement, 25, 29, 35n.
  scintillation method, 58
  transmutation, 30, 34, 35, 55

Radius
  atoms, 55, 56, 70
  emulsion grains, 47
Rayleigh, Lord, 10, 52
Regener, Erich, 58
Richardson, O. W., 80, 83
Rutherford, Ernest, Baron, 24 ff., 55, 57 ff.

Saturation photocurrent, 85
Scattering
  α particles, 57 ff.
  electromagnetic radiation, 97 ff.
  electrons by solids, 106
Soddy, Frederick, 24 ff., 53n., 54b
Sommerfeld, Arnold, 88
Stefan-Boltzman law, 8, 12, 14
Stern, Otto, 89 ff., 118 ff.

Thomson, Sir George, 117, 134n., 135b
Thomson, Sir J. J., 55, 98, 117
Thomson model of atom, 55, 61, 62, 67b
Transmutation (*see under* Radioactivity)

Ultraviolet catastrophe, 11

Wien, Willy, 8, 21b
Wien's displacement laws (*see also under* Cavity radiation), 9, 14, 19, 21b